OECD Green Growth Studies

Improving Energy Efficiency in the Agro-food Chain

This work is published under the responsibility of the Secretary-General of the OECD. The opinions expressed and arguments employed herein do not necessarily reflect the official views of OECD member countries.

This document and any map included herein are without prejudice to the status of or sovereignty over any territory, to the delimitation of international frontiers and boundaries and to the name of any territory, city or area.

Please cite this publication as:
OECD (2017), *Improving Energy Efficiency in the Agro-food Chain*, OECD Green Growth Studies, OECD Publishing, Paris.
http://dx.doi.org/10.1787/9789264278530-en

ISBN 978-92-64-27852-3 (print)
ISBN 978-92-64-27853-0 (PDF)

Series: OECD Green Growth Studies
ISSN 2222-9515 (print)
ISSN 2222-9523 (online)

The statistical data for Israel are supplied by and under the responsibility of the relevant Israeli authorities. The use of such data by the OECD is without prejudice to the status of the Golan Heights, East Jerusalem and Israeli settlements in the West Bank under the terms of international law.

Photo credits: Cover © design by advitam for the OECD

Corrigenda to OECD publications may be found on line at: *www.oecd.org/about/publishing/corrigenda.htm*.

Foreword

In order to achieve sustainable development and meet their green growth objectives, governments need to facilitate the efficient use of natural resources. In many countries, the agro-food sector is a significant consumer of fossil fuel-based energy. Therefore, improving the energy-use efficiency of the agro-food chain should be a key priority and a core element of green growth strategies.

A range of factors – including economic growth, rising consumer incomes, and increasing demand for convenience products – all have the effect of increasing energy usage by the sector. The amount of energy used is forecast to rise in the coming years. Moreover, high dependence on fossil fuels means that the agro-food sector is also associated with large greenhouse gas emissions. It is therefore becoming increasingly important to consider the ways in which the food supply chain can improve its energy efficiency.

This report explores how this challenge – that of increasing output per unit of energy used – can be attained, and suggests a set of policy recommendations that governments can introduce to achieve this end. Measures are considered in relation to both producers and consumers.

The Executive Summary provides a synthesis of the key findings of the report. Chapter 1 explains why the issue of energy efficiency in the agro-food sector requires attention. Chapter 2 looks at the ways in which energy is used on-farm, and explores the possibilities for achieving improved energy use, while Chapter 3 examines energy usage and efficiency with regard to the off-farm part of the food supply chain. Chapter 4 focuses on measures that could be enacted across the supply chain, and assesses the role played by consumers. Chapter 5 concludes the report by presenting a range of measures that could be taken to improve energy efficiency throughout the agro-food chain.

This report is part of broader OECD work on green growth in agriculture, further information on which can be found at: www.oecd.org/tad/sustainable-agriculture/greengrowthforfoodagricultureandfisheries.htm

Acknowledgements

Dimitris Diakosavvas was the project leader and the main author of this report. The report draws on background material prepared by three consultants: Clara Thompson-Lipponen (OECD), Bruce Traill (Emeritus Professor, University of Reading, United Kingdom) and Dominic Moran (Scottish Agricultural College, United Kingdom). Comments from delegates of the JWPAE, and OECD and International Energy Agency (IEA) colleagues were very much appreciated, in particular those from Eric Masanet, Adam Brown, Ronald Steenblik and Céline Giner. Thanks also to Mark Mateo for statistical support. Editorial assistance was provided by Robert Akam. Theresa Poincet provided invaluable secretarial assistance throughout the production process and prepared the report for publication. Michèle Patterson co-ordinated its production.

This report was prepared by the OECD Trade and Agriculture Directorate. It was declassified by the OECD Joint Working Party on Agriculture and the Environment (JWPAE) in April 2017.

Table of contents

Tables

Figures

Acronyms and abbreviations

ACEE	American Council for an Energy-Efficient Economy
ACNV	Automatically Controlled Natural Ventilation
ADEM	Agence de l'Environnement et la Maîtrise de l'Énergie (France)
BIAC	Business and Industry Advisory Committee to the OECD
bBtu	Billion British thermal unit
Btu	British thermal unit
CCAs	Climate Change Agreements (United Kingdom)
CEP	Clean and Efficient Programme (The Netherlands)
CEFC	Clean Energy Finance Corporation (Australia)
CHP	Combined Heat and Power
CO_2	Carbon dioxide
CO_{2e}	Carbon dioxide equivalents
DK	Danish krone
DEFRA	Department for Environment, Food and Rural Affairs
EECA	Energy Efficiency and Conservation Authority (New Zealand)
EED	Energy Efficiency Directive (European Union)
EJ	Exajoule
EPA	Environmental Protection Agency (United States)
EQIP	Environmental Quality Incentive Program (United States)
EU	European Union
EUR	Euro
ETS	Union Emissions Trading Scheme
FAO	Food and Agriculture Organization of the United Nations
FDF	Food and Drink Federation
GBP	Great Britain Pounds
GHG	Greenhouse gas
Gj	Giga joule (= 10^9 J)
Gt	Gigatonnes
GW	Gigawatts
ha	Hectares
HVAC	Heating Ventilation and Air Conditioning
IFA	International Fertilizer Industry Association
IEA	International Energy Agency
IFES	Integrated Food-Energy System
IPCC	Intergovernmental Panel on Climate Change
ISIC	International Standard Industrial Classification
J	Joule

Kg	Kilogramme
Kgoe	Kilogramme of oil equivalent
Ktoe	Kilotonne of oil equivalent
kVA	kilovolt-ampere (1000 volt amperes)
kWh	Kilowatt-hour
$kWhyr^{1}hd^{-1}$	kWh per year per animal
L	Litre
LCA	Life Cycle Assessment
LPG	Liquefied petroleum gas
LTAs	Long Term Agreements
LU	Livestock unit
MACC	Marginal Abatement Cost Curve
MBtu	Mega Btu
Mtoe	Million tonnes of oil equivalent
MJ	Mega joule = 10^{6} J
Mt	Metric tonnes
MWe	Megawatt electrical
N_2O	Nitrous oxide
NH_4	Methane
NH_3	Ammonia
NFU	National Farmers' Union (England)
NRDC	Natural Resources Defence Council
OECD	Organisation for Economic Co-operation and Development
OFAG	Office fédérale de l'agriculture (Switzerland)
PJ	Peta joule (= 10^{15}J)
qBtu	Quadrillion Btu
R&D	Research and Development
REAP	Rural Energy for America Program
SMEs	Small and medium-sized enterprises
t	Metric ton
tCO_2	Tonnes of CO_2
Tj	Terajoules
TPES	Total primary energy supply
Toe	Ton of oil equivalent
TWh	Terawatt-hours
UNFCCC	UN Framework Convention on Climate Change
USD	United States Dollar
WRAP	Waste and Resources Action Programme (United Kingdom)

Executive summary

Almost all activities in the food system depend on some form of energy, which is currently mainly provided by fossil fuels. The need to use scarce natural resources efficiently, reduce greenhouse gas emissions, minimise energy costs and foster the competitiveness of the agro-food sector highlights the importance of the energy efficiency issue: using less energy to provide the same level of output and services. Improving the energy-use efficiency of the agro-food chain is a key priority in several OECD countries and a core element of green growth strategies.

The current food chain system is energy-intensive, highly dependent on fossil fuels

- Available empirical studies – mainly on EU countries and the United States – suggest that the food system accounts for as much as 20% of total energy use in some OECD countries. At the farm level, energy is consumed both directly – as fuel or electricity to power farm activities – and indirectly – in the form of fertilisers and chemicals produced off-farm. In the OECD area, on average, direct energy use by agriculture represents only 2% of total energy consumption. Moreover, energy accounts for an important and highly variable share of food costs.
- Much of the growth in energy use has been driven by changes in lifestyle and consumer preferences, particularly the demand for more processed and ready-to-eat foods. There is also a substantial amount of food wastage throughout the system, which reduces energy efficiency.
- Food production systems vary substantially in their energy use and potential for energy efficiency, depending mainly on the particular activity involved, combined with agro-ecological conditions. This means that opportunities for energy savings are numerous; however, at this stage, identifying opportunities requires further research.
- It is widely recognised that energy efficiency gains can deliver private benefits. The private sector is already taking action in response to government initiatives and legislation, rising energy prices and the desire of many companies to improve social and environmental performance.
- In the OECD area, improvements in energy efficiency have been achieved primarily through the application of proven energy conservation management practices and technologies, and projects and activities that can lead to a quick return on investment. Measures include nutrient management, precision farming, waste recovery, and more efficient manufacturing, refrigeration and transportation technologies. Some interventions have the benefit of lowering energy-use costs, while also reducing the GHG emissions associated with inputs.
- Crucial elements in delivering energy efficiency gains in the private sector are: i) having the infrastructure for energy management that incorporate clearly defined responsibilities, robust data, on-going monitoring and performance review; and ii) having targets relating to the development of innovations and to the wider deployment of innovative technologies.
- Despite existing efforts, market failures, policy-induced market distortions, and financial, organisational and behavioural barriers all combine to impede private sector's energy-efficiency initiatives. Food businesses are calling for a clear, consistent, regulatory environment that supports energy-efficient gains, and within which the private sector can thrive. Overall, OECD governments are becoming increasingly aware of the need to improve energy efficiency through addressing policy failures and by encouraging public-private partnerships.

Key recommendations

Foster energy efficiency in the food chain, ensuring coherence with other policy goals

- Transmitting appropriate price signals to producers and consumers is essential if greater energy efficiency across the food chain is to be achieved. Explicit or implicit subsidies that distort price signals need to be eliminated if firms are to be induced to make greener decisions on resource use. Likewise, the low tax rates and exemptions on fuel used in agriculture suggest that some of the lowest-cost opportunities to reduce carbon emissions are being forgone. In many countries, a reappraisal may be warranted to explicitly determine whether current tax settings are appropriately adapted to their environments, social and economic goals. Market mechanisms must also be allowed to operate in order to provide incentives for the private sector to develop and adopt new technologies and innovations. Public-private partnerships could have an important role to play.
- Policies aimed at improving energy efficiency must also be coherent and consider the synergies and trade-offs with policies addressing such issues as productivity, water use, and health and food safety. For example, methods used to reduce energy inputs that also lower productivity, such as reducing (rather than optimising) the amount of fertiliser applied, are rarely beneficial to the farmer. Likewise, there may be a case for increasing energy inputs over time, in order to improve productivity and water-use efficiency.
- A best-practice strategy should:
  - identify the barriers to cost-effective efficiency investments – and attempt to overcome them;
  - assess opportunities for energy efficiency improvements, and prioritise action in those areas of the agro-food chain where government policies are most likely to yield the largest, most cost-effective improvements;
  - set clear objectives and timelines, and establish evaluation methods; and
  - ensure coherence with energy, environmental/climate and economic strategies.

Improved public awareness of the energy used to produce food

- The effectiveness of price signal measures can be enhanced by complementary information measures, although these measures would need to be part of a broader policy package.
- Due consideration should be given to the relative cost-effectiveness and challenges of changing production behaviour versus that of changing the consumption side. Public information campaigns, education and labelling to enable consumers to make informed choices and reduce waste, are necessary but have long-term, often unforeseeable effects. If governments work together with industry to promote reformulation, reduce packaging and deliver less carbon-intensive foods, this may produce an effective result, although fiscal measures may also be necessary to incentivise change.

Encourage greater understanding of where efficiencies can be made

- As a result of data and methodological deficiencies, and the diversity of approaches to measure energy consumption and efficiency of the agro-food chain, considerable uncertainty continues to exist concerning the measurement and monitoring of the energy levels used and the energy efficiency of the various components of the food chain. Efforts are necessary to design a common methodology to improve data collection and obtain a clearer understanding of the energy efficiency potential of the components of various food production systems, leading to the creation of more effective policies and business responses.

Chapter 1

Why does energy efficiency in the agro-food chain matter?

Energy is crucial for economic growth and a critical component in the ability of the agro-food sector to improve productivity, competitiveness and sustainability. Improving the efficiency of energy use – using less energy to provide the same level of output and service – is an important tool that policy makers can use to ensure a number of positive outcomes that can deliver several government priorities, from economic growth to greenhouse gas reduction to energy security and food security. To set the scene this Chapter discusses the increasing demands for energy throughout the food chain. The synergy and overlapping policy goals between energy efficiency, food waste and GHG emissions mitigation agendas are noted, and conceptual and methodological issues involved in measuring energy use and efficiency are highlighted.

The efficient use of natural resources has become a key priority for policy making in OECD countries and a core element of green growth strategies. But increasing efficiency in the use of natural resources such as land, water and energy, while meeting the demands of an expanding and wealthier global population, is challenging (OECD, 2013).

This report examines the issue of energy use and efficiency across the whole food chain, focusing on private-sector initiatives and the role that government can play in unlocking greater energy efficiency within the sector. Energy demand across the sector is projected to grow steadily, both in agriculture and further downstream in the manufacturing and distribution processes.

Increasing dependence on energy usage (mainly fossil fuels) throughout the entire food chain raises concerns about the impact of high or variable energy prices on production costs, competitiveness, the final price of food for the consumer, as well as concerns about energy security. In addition to these concerns, the use of energy in the food chain can also have environmental impacts, such as greenhouse gas (GHG) emissions. While progress has been made, the private and government sectors can do more to ensure that the full energy efficiency potential of the food system is materialised.

The report synthesises existing literature on energy use in the food system to explore actions that have – or could be – undertaken at the various stages in the food chain: however, detailed technical innovations are industry- and firm-specific and are not addressed here. Issues explored in the report mainly concern areas where public-sector efforts to create an environment in which the private sector is enabled to realise its full potential for increasing the productivity of the food chain, in a sustainable manner.

The challenge of increasing energy usage across the agri-food sector

Food supply or value chains are diverse and highly complex with many final products drawing on inputs from a variety of industries. The type of and the way energy is used in the food chain can influence the extent to which the food system in a country will be able to support growth and productivity objectives in an environmentally sustainable manner. This is the case as energy is used throughout the entire food chain: in the production of crops, fish, livestock, and forestry products; in post-harvest operations; in food storage and processing; in food transport and distribution; and in food preparation. Moreover, energy contributes to the transformation and reuse of the various forms of by products and waste which the food production process generates.

Food is a composite and perishable product and the amount of energy for bringing it "from farm to fork" varies greatly from one product to another. Even when considering the same type of product, the energy "cost" differs notably, reflecting changes on cultivation area, farming practices, efficiency of processing and storage, season of production and/or consumption, transportation needs, etc. Moreover, in most OECD countries, supply chains have evolved in multi-stage production and processing operations with varying degrees of vertical and horizontal integration. In addition, the diversity of product-specific chains means that a precise accounting of energy consumed in food production is extremely challenging (Annex 1A.1).

Energy use at any stage of the food chain may comprise direct use by a specific production process, plus indirect use where a stage uses energy ingredients from other industries that contribute to a final product (Box 1.1). Some parts of the supply chain are direct energy users, which makes measuring monitoring of energy efficiency more straightforward. Accounting for and reducing indirect energy use is more challenging, especially in the case of farming.

Box 1.1. Defining energy usage in the food supply

Energy efficiency is the use of a lower amount of energy to provide the same level of output and services. This can be defined in a pure sense as the reduced use of all energies – fossil fuels and renewables alike – by means of energy conservation measures. It is important to distinguish "energy efficiency", however, from energy conservation, which is a broader term which can also include forgoing a service rather than changing the efficiency with which it is provided.

Most food products have a specific life-cycle which involves a combination of direct energy use at specific stages of the cycle (e.g. crop drying on farm) plus indirect energy use related to the production of other intermediate inputs that ultimately go into a final product. The more complex a product life-cycle, the less adequate direct measures become in measuring total energy consumption. For example, with regard to the agricultural sector, direct energy for farming would include electricity, heating fuel and machinery fuel used in crop production, grain drying, animal and animal production, heating/cooling of animal houses, transportation of farm products and personal energy use (for example, heating farm-house and driving to town). Indirect energy would consist of the energy consumed in the production, packaging and transport to the farm-gate of fertilisers, pesticides, farm machinery and buildings (CAEEDAC, 2000).

In OECD countries, agriculture's direct use of energy is roughly in line with its contribution to GDP– around 2%, but the food system – both upstream and downstream – as whole is a much more substantial contributor to energy use (Box 1.2). For example, a study by the FAO (2011) suggests that, globally, the food system accounts for around 30% of the world's total end-use energy consumption – with more than 70% consumed beyond the farm gate – and produces about one-fifth of the world's GHG emissions. In addition, more than one-third of the food produced is lost or wasted, and with it about 38% of the energy consumed in the food chain. Moreover, in all components of the entire food chain, most of energy used originates from fossil fuels.

Energy use, including in the agri-food sector, is projected to increase significantly in the coming decades (OECD, 2012). Economic growth, rising consumer incomes, changing technologies, demographics and changes in lifestyles all contributing to altered patterns of food consumption and production, and demand for energy.

Box 1.2. The food chain is an important energy consumer

Consistent comparison of the total energy (direct and indirect) consumed by the food chain and its components is hampered by the use of different definitions and boundaries for supply chains. But figures suggest that this could be as high as 19% of total energy for the United States (Pimental, 2006; Canning et al., 2010; Schnepf, 2004). For the EU countries, studies report shares of 17% for the EU27 (Monforti-Ferrario et al., 2015); 14% for France INSEE (2015); 13% for Sweden (Wallgren and Hojer, 2009); and 18% for the United Kingdom (Tassou, et al., 2014). In France, the agro-food sector is the third most important consumer of energy in the country.

Somewhat surprisingly, the recent and very comprehensive study for the EU27, which is based on LCA methodology, finds that agriculture (crop cultivation and animal rearing) was the most energy-intense phase of the food system in 2013 – accounting for nearly one-third of the total energy consumed in the food production chain (Monforti-Ferrario et al., 2015). The second most important phase of the food life cycle was industrial processing – accounting for 28% of total energy use. About 60% of the energy embedded (the sum of all energy inputs along the production chain) in the food consumed in the EU27 was derived from agriculture and logistics – two sectors that are largely dominated by fossil fuel use, in which the penetration of renewable energies is still relatively small.

FAO's (2011) indicative estimates on energy consumed by the agri-food sector and its components show that: i) globally, the agri-food sector accounts, directly and indirectly, for around 30% of the world's total end-use energy consumption – with more than 70% consumed beyond the farm gate; ii) the food chain produces about one-fifth of the world's GHG emissions; iii) primary agriculture and fishery production accounts for around one-fifth of total energy demand, but produces two-thirds of the GHGs; iv) high-income countries use a greater proportion of this energy for processing and transport, while in low-income countries cooking consumes the highest share; and v) more than one-third of the food produced is lost or wasted, and with it about 38% of the energy consumed in the food chain; and vi) the share of energy used for retail, preparation and cooking is considerably higher in low-GDP countries (about 45%) than in high-GDP countries (30%).

In the European Union, in all the steps of the entire food life cycle chain, most of energy used originates from fossil fuels (79%), followed by nuclear energy (14%) (Monforti-Ferrario et al., 2015). Hydro energy plays an important role in industrial processing, while the energy from biomass is significant in the end-of-life stage. While EU countries have made important progress in incorporating renewable energy across the economies, the share of renewables in the food system remains relatively small, accounting for only 7% of the total energy consumed by the food life cycle.

A number of studies have addressed food-related energy use in the United States. These studies generally indicate that: i) food-related energy use continues to consume a substantial share of the total national energy budget; ii) food-related energy use of households has been the largest among the seven supply chain stages considered (agriculture, processing, packaging, transportation, wholesale and retail, food services, households); iii) food processing shows the largest growth in energy use as both households and food-service establishments increasingly outsourced manual food preparation and clean-up to manufacturers; and iii) food-related energy flows may have increased significantly over time (Canning et al., 2010; 2017). Population growth, higher per capita food expenditures and greater reliance on energy-using technologies boosted food-related energy consumption.

In the United Kingdom, the agri-food sector is highly dependent on energy, principally oil and gas, and no part of the supply-chain is immune from either increases in the cost of energy or disruptions to supply. Given the length and complexity of the supply chain, different stages are vulnerable at different times, depending on energy type, with potentially different consequences. Domestic operations, cooking and refrigeration make the highest contribution, followed by manufacturing, commercial transport, agriculture, retail and catering (DEFRA, 2013). Energy use in food production (primary production, processing and manufacturing) is very diverse and varies with product. The review of life cycle assessment studies undertaken by DEFRA, 2013) showed that energy used in production, expressed as MJ/kg, ranged between 2.2 (potato) and 51.3 (cheese). Energy use within the life cycle of products also varied greatly; primary production accounted for between 17% (yoghurt) and 63% (milk). Although data availability on energy usage within processing was sparse, the range is still considerable, being between 3% (rice and onion) to 64% (bread). Energy use within the transport phase was influenced by bulky items (apple, 40%) and distance (rice, 28%).

The principal reason for the expanded use of energy in the agro-food sector is the search for convenience by consumers, in particular by consuming more prepared foods and a larger amount of food outside the home, both of which use more energy (Blandford, 2013). Changes in food consumption patterns are affecting a large number of countries, particularly emerging economies. While future increases in energy demand generated by agriculture *per se* may be relatively modest, further changes in lifestyles and food consumption patterns across the world could impose greater demands on global energy supplies.

The growth in the production of "convenience" foods and changes in the presentation of foods to consumers (e.g. sales of washed and packaged vegetables rather than in their relatively unprocessed state) could not only increase energy usage, but could also generate a higher waste stream in the form of packaging. The standards set by retailers (e.g. requirements on the size and appearance of fruit and vegetables) can also add to the amount of material entering the waste stream as products that do not meet those requirements are unable to find a market. Much of the food product waste which used to be fed to livestock now ends up in landfill sites.

This trend comes along with additional energy requirements associated with climate change impacts. In the future, more extreme weather, along with a greater number of dry periods in spring and summer and more wet years, could lead to increases in energy use. In dry conditions, additional energy would be needed to irrigate crops, and in wet conditions, machines and tractors would use more fuel.[1] OECD work on fossil fuels suggests that although energy is essential for growth, current energy use patterns in the form of fossil fuels are unsustainable if climate change targets are to be met (OECD, 2015a).

Agriculture also causes CO_2 emissions by using energy (e.g. fuel, electricity, heating) and is the final user of several inputs that are produced in an energy-intensive manner (e.g. fertilisers, pesticides).[2] In the developed world, beyond the farm gate, CO_2 emissions are highest from food manufacturing, transport and home related activities (e.g. cooking, refrigeration) and significant but slightly lower from each of packaging, retail and catering (FAO, 2011; Sims et al., 2015). Refrigeration is a major source of GHG emissions in manufacture, retailing and catering and the ubiquity of refrigeration systems has fostered the development of chilled foods. These patterns, well established in the developed world, are a major tendency of food systems in emerging economies.

Emissions from on-farm energy use and production of fertilisers account for approximately 8 to 10% of global agricultural emissions (Sims et al., 2015; FAO, 2011; Wirsenius et al., 2011). One study estimates that in the absence of abatement measures, annual global emissions of GHG from agriculture are likely to increase by 30% by 2030 when compared to estimated levels in 2005 (McKinsey & Company, cited in Wreford et al., 2010). In the United States, fossil fuel use linked to domestic food consumption accounted for 13.6% of economy-wide CO_2 emissions from fossil fuels in 2007 (Canning, et al., 2017).

Moreover, one important characteristic of agricultural emissions is that they result from the activities of a large number of small-scale emitters, which together with the diversity in the conditions of production within and across countries leads to large heterogeneities in abatement costs. Such heterogeneities have important consequences on the design of cost-effective mitigation policies (De Cara and Jayet, 2011).

Improving energy efficiency in agri-food supports a range of energy-related goals

Improving the efficiency of energy use – using less energy to provide the same level of output and service – is widely recognised by many governments around the world as the most cost-effective and readily available means to address numerous energy-related issues, including energy security, the social and economic impacts of high energy prices and concerns about climate change (IEA, 2014a).[3] At the same time, energy efficiency increases business competitiveness and promotes consumer welfare.

Successful energy efficiency projects can bring additional multiple advantages which extend far beyond the reduction of energy bills or emissions. Several authors have found that technologies which increase energy efficiency can also bring improvements to the production process, such as lower operational and maintenance costs, increased production yield, open outlets in new food markets that require certification of sustainability or energy performance and safer working conditions, all of which increase the productivity, overall efficiency and profitability of a firm (Worrell et al., 2001; IEA, 2014a; OECD, 2015b).

Energy efficiency measures have enabled reductions in capital and labour costs which have been even greater than the energy savings themselves (Worrell et al., 2001). Indeed, according to the IEA (2014a), the value of the productivity and operational benefits derived from energy efficiency can be up to 2.5 times the value of energy savings, depending on the value and context of the investment. Food industry projects which were initially calculated to have a three- to four-year payback have therefore been found to deliver the full return on investment in just one year when benefits such as these are integrated into the overall assessment. Energy efficiency improvement can also enable companies to increase earnings in times of high energy price volatility, as well as to improve consumer perception of their product and thereby increase sales (Verghese et al., 2012).

Beyond economic benefits, energy efficiency technologies have also been found in some cases to have less obvious but equally important effects, such as water savings and waste minimisation, which can in turn enable energy savings in water and waste management (Charpentier, 2016; Worrell et al., 2001).

A perspective of where emissions are greatest and where savings are most cost-effective is useful for understanding where energy efficiencies in the food chain can be made. This is important because often commonly held beliefs are counter-intuitive, for example the view that local foods mean lower energy use (and emission) emissions (Van Hauwermeiren et al., 2007). For example, comparison of energy and emissions based on simple concepts such as Food Miles – the distance that a product travels can be misleading because of major differences in emissions in production. Products may differ substantially in the energy required to produce them, for example, and a product that has travelled a long distance may have a lower emissions content than one produced locally under energy-intensive conditions.

Private sector efforts are increasingly focused on energy efficiency

The private sector is paying increased attention towards the importance of energy efficiency due to growing awareness of the limitations of fossil fuels for sustainable productivity, of the implications of GHG emissions and of variable and rising energy prices on competitiveness and productivity. Indeed, energy efficiency is increasingly being recognised as one of the most important and cost-effective solutions to reduce GHG emissions and other important air pollutants (IEA, 2014a; Masanet et al., 2012).

Higher energy prices necessarily stimulate the search for greater efficiency in energy use in the food chain. Increases in the price of energy, for example, would have an impact both through the direct effect on energy used on farms and through its effect on the prices of agro-chemicals and other services such as transportation. However, farmers can be adept at economising on the use of inputs in response to higher prices.

When increases in energy and fertiliser prices become significant for farm profitability, for example, farmers would search in adopting management practices to reduce their usage. Lower usage of energy could be achieved through such measures as employing machinery less intensively and servicing engines more frequently; lower consumption of fertiliser could be achieved through the greater use of soil testing, changes in plant populations and the adoption of precision application methods. It will therefore be a natural tendency for firms in the agro-food chain to search for ways to economise in the use of energy if its price continues to increase in real terms. Recent low fossil fuel prices could, however, discourage efforts for further improving energy efficiency. Nevertheless, the fall in oil prices has also provided favourable conditions in some countries to reduce end-use fossil fuel subsidies, which undermine the economic attractiveness of energy efficiency investments (IEA, 2015).

Important progress has been made by the private sector to improve energy efficiency in the food chain through innovation, investments in more efficient technologies and adoption of more energy-efficient management practices. But more is anticipated, and more can still be done. For example, the **European Union** has launched a major policy initiative on energy efficiency, within which it identifies the potential to reduce energy use by 20% through cost-effective demand- and supply-side measures.

Governments have a role to play in encouraging energy efficiency

Capturing economic incentives for efficiency improvements requires, *inter alia*, transparent energy pricing, information on opportunities to improve efficiency and investments in research and development. In these areas, governments have an important role to play.

Governments can play a decisive role in boosting energy efficiency in the food chain by putting in place appropriate business-enabling policies, in order to allow the private sector to realise its full potential. The key drivers of energy efficiency – namely investment and innovation – require the creation of enabling policy frameworks in which private sector-led and collaborative investment and innovation initiatives can thrive. This requires an approach in which policy coherence and partnerships with the private sector are key aspects.

The private sector response to the challenges of improving energy efficiency can be enhanced if firms are able to benefit from the business opportunities that this can create. However, there are many challenges to improving energy efficiency in the food chain that can be identified. They include, among others, market distortions, lack of information, co-ordination and risk aversion elements. Energy efficiency in the agro-food chain also has to be considered alongside other factors that drive investment decisions such as new product development, market growth and production location decisions to meet those markets.

In addition, firms are likely to intensify their efforts to increase energy efficiency of food products if consumers respond by purchasing energy-efficient products. Greater government-led awareness raising campaigns and knowledge sharing in all parts of the agro-food chain – including consumers, as well as the role of regulation and explicit or implicit taxes to internalise the external environmental effects of production and consumption decisions, are very important.

Policy makers must bear in mind that improving energy efficiency may save less energy than expected due to a "rebound" of energy use and may, in some cases, actually lead to an increase in energy use. As energy consumers save on energy cost through energy efficiency, they may spend their savings on other energy-intensive activities, or increase their demand for the new service, thereby countering the potential savings of energy. For example, consumers buying more energy-efficient household appliances may then use them more frequently because they are cheaper to run (Gillingham, Rapson and Wagner, 2015; Sorrell and Dimitropoulos, 2008; Tollefson, 2011).[4]

Further, policy makers must take into account that improving energy efficiency does not necessarily translate into reduced CO_2 emissions: the savings depend on the type of energy. If the energy is supplied from fossil fuels, then improved efficiency will cut emissions. But if the energy is supplied by a low-carbon source such as renewables, then improving efficiency may have little impact on emissions.[5] Nonetheless, improving energy efficiency overall is a key tool for reducing CO_2 emissions, alongside energy conservation and low-carbon energy sources such as renewables, carbon capture and storage.

Additionally, if reducing energy use is the policy goal, then energy efficiency is only one of a number of factors that impact energy use, and energy conservation may or may not be associated with an increase in energy efficiency – depending on the input-output relationship. This is, energy consumption may be reduced with or without an increase in energy efficiency, and energy consumption may increase alongside an increase in energy efficiency.

Notes

1. At the same time, increasing temperature may reduce the need for heating in many places as well.

2. These emissions could be eliminated where they stem from electricity generated from fossil fuels, if zero-emissions electricity sources were to replace current infrastructure.

3. Improving energy efficiency in most of the food chain will have the effect of reducing GHG emissions. Agriculture, however, is the exception because much of the sector's GHG emissions take the form of nitrous oxide (N_2O) and methane (NH_4).

4. The rebound effect is the reduction in expected gains from new technologies that increase the efficiency of resource use, because of behavioural or other systemic responses. The rebound effect is generally expressed as a ratio of the lost benefit compared to the expected environmental benefit when holding consumption constant. There are different rebound effect types, depending on the magnitude of the rebound effect. When the actual resource savings are negative because usage increased beyond potential savings (the rebound effect is higher than 100%), this situation is commonly known as the Jevons paradox (Gillingham, Rapson and Wagner, 2015; Sorrell and Dimitropoulos, 2008).

5. For example, when comparing electric and non-electric appliances, it is important to consider the efficiency of the power generation as well: switching from a 90% efficient gas boiler to a "100% efficient" electric heater might increase energy use and emissions if the electricity comes from fossil-fuel power plants, which themselves are inefficient, as they lose much energy as waste heat.

Bibliography

Blandford, D. (2013), *Green Growth in the Agro-food Chain: What Role for the Private Sector?,* Background Paper for OECD/BIAC Workshop "Green Growth in the Agro-food Chain: What Role for the Private Sector?" , 24 April 2013, www.oecd.org/tad/sustainable-agriculture/Session%201%20BLANDFORD_FORM.pdf.

Burney, J. (2001), "How much energy does it take to make your meal? Understanding the energy inputs into the food system at different scales", Programme on Food Security and the Environment, TomKat Center for Sustainable Energy Stanford University, https://fsi.stanford.edu/sites/default/files/jennifer_burney.pdf.

Canadian Agricultural Energy End-Use Data Analysis Center (CAEEDAC) (2000), *Direct Energy Use in Agriculture and the Food Sectors – Separation by Farm Type and Location*, A Report to Natural Resources Canada, by The Canadian Agricultural Energy End Use Data and Analysis Centre final report, www.usask.ca/agriculture/caedac/pubs/Food.PDF

Canning, P., S. Rehkamp, A. Waters and H. Etemadnia (2017), *The Role of Fossil Fuels in the U.S. Food System and the American Diet*, Economic Research Service, United States Department of Agriculture, ERS Report No. 224.

Canning, P., A. Charles, S. Huang, R. Polenske and A. Waters (2010), *Energy Use in the U.S. Food System* Economic Research Service, United States Department of Agriculture, ERS Report No. 94.

Carlsson-Kanyama, A., M. Ekström and H. Shanahan (2003), "Food and life-cycle energy inputs: consequences of diet and ways to increase efficiency", *Ecological Economics*, Vol. 44, Issues 2-3.

Charpentier, O. (2016), *Efficient Energy Use in Agri-food Chains*, Inter-American Institute for Co-operation on Agriculture (IICA), www.iica.int/sites/default/files/publications/files/2016/B3984i.pdf.

De Cara, S. and P.A. Jayet (2011), "Marginal abatement costs of greenhouse gas emissions from European agriculture, cost effectiveness, and the EU non-ETS burden sharing agreement", *Ecological Economics*, Vol. 70, pp. 1680-1690.

De Vries, M. and I. de Boer (2010), "Comparing environmental impacts for livestock products: A review of life-cycle assessments", *Livestock Science*, Vol. 128.

Department for Environment, Food and Rural Affairs (DEFRA) (2013), *Energy Dependency and Food Chain Security*, Report FO0415, London, United Kingdom, http://randd.defra.gov.uk/Default.aspx?Menu=Menu&Module=More&Location=None&Completed=0&ProjectID=16433.

EUROSTAT (2012), *Agri-environmental indicator – energy use*, ec.europa.eu/eurostat/statistics-explained/index.php/Agri-environmental_indicator_-_energy_use

Food and Agriculture Organization of the United Nations (FAO) (2011), *Energy Smart Food for People and Climate – Issue Paper*, Rome, www.fao.org/docrep/014/i2454e/i2454e00.pdf.

Gillingham, K., D. Rapson and G. Wagner (2015), "The Rebound Effect and Energy Efficiency Policy", *Review of Environmental Economics and Policy*, Vol. 10, No. 1.

Heller, M. and G. Keoleian (2000), "Life Cycle-Based Sustainability Indicators for Assessment of the U.S.", *Food System*, The Center for Sustainable Systems, Michigan, United States.

International Energy Agency (IEA) (2015), *Energy Efficiency Market Report*, www.iea.org/publications/freepublications/publication/energy-efficiency-market-report-2015-.html, International Energy Agency, Paris.

IEA (2014a), *Capturing the Multiple Benefits of Energy Efficiency*, www.iea.org/publications/freepublications/publication/Captur_the_MultiplBenef_ofEnergyEficiency.pdf, International Energy Agency, Paris.

IEA (2014b), *Energy Efficiency Indicators: Essentials for Policy Making*, www.iea.org/publications/freepublications/publication/energy-efficiency-indicators-essentials-for-policy-making.html, International Energy Agency, Paris.

Institut national de la statistique et des études économiques (INSEE) (2015), *Les consommations d'énergie dans l'industrie en 2013*, http://atee.fr/management-de-lenergie-statistiques-etudes/infographie-consommation-denergie-de-lindustrie-en-france

Masanet, E., P. Therkelsen and E. Worrell (2012), *Energy Efficiency Improvement and Cost Saving Opportunities for the Baking Industry*, Ernest Orlando Lawrence Berkeley National Laboratory, www.energystar.gov/sites/default/files/buildings/tools/Baking_Guide.pdf.

Monforti-Ferrario, F., J.-F. Dallemand, I. Pinedo Pascua, V. Motola, M. Banja, N. Scarlat, H. Medarac, L. Castellazzi, N. Labanca, P. Bertoldi, D. Pennington, M. Goralczyk, E. M. Schau, E. Saouter, S. Sala, B. Notarnicola, G. Tassielli and P. Renzulli (2015), *Energy use in the EU food sector: State of play and opportunities for improvement*, European Commission, Joint Research Centre, Brussels, Belgium.

Office fédérale de l'agriculture (OFAG) (2015), *Rapport agricole,* Bern, Switzerland, http://2015.agrarbericht.ch/fr.

OECD (2015a), *OECD Companion to the Inventory of Support Measures for Fossil Fuels 2015*, OECD Publishing, Paris, DOI: http://dx.doi.org/10.1787/9789264239616-en.

OECD (2015b), *An Introduction to Energy Management Systems: Energy Savings and Increased Industrial Productivity for the Iron and Steel Sector*, OECD Publishing, Paris, www.oecd.org/sti/ind/Energy-efficiency-steel-sector-1.pdf.

OECD (2013), *Policy Instruments to Support Green Growth in Agriculture: A Synthesis of Country Experiences*, OECD Publishing, DOI: http://dx.doi.org/10.1787/9789264203525-en.

OECD (2012), *OECD Environmental Outlook to 2050: The Consequences of Inaction*, OECD Publishing, Paris, DOI:10.1787/9789264122246-en.

Pimentel, D. (2006), *Impacts of organic farming on the efficiency of energy use in agriculture*, The organic Centre, Cornell University.

Saunders, C. and A. Barber (2008), "Carbon Footprints, Life Cycle Analysis, Food Miles: Global Trends and Market Issues", Political Science, Vol. 60, No. 1.

Schnepf, R. (2004), *Energy Use in Agriculture: Background and Issues*, CRS Report for Congress, http://nationalaglawcenter.org/wp-content/uploads/assets/crs/RL32677.pdf.

Sims, R., A. Flammini, M. Puri and S. Bracco (2015), *Opportunities for Agri-Food Chains to Become Energy-Smart*, www.fao.org/3/a-i5125e.pdf.

Sorrell, S. and J. Dimitropoulos (2008), "The rebound effect: Microeconomic definitions, limitations and extensions", *Ecological Economics*, Vol.65, pp. 636-649.

Tassou, S., M. Kolokotroni, B. Gowreesunker, V. Stojkeska, A. Azapagic, P. Fryer and S. Bakalis (2014), "Energy Demand and Reduction Opportunities in the UK Food Chain", *Energy*, Vol. 167, No. 3.

Tollefson, J. (2011), "Experts tangle over energy-efficiency 'rebound' effect", *Nature.*

Van Hauwermeiren, A., H. Coene, G. Engelen and E. Mathijs (2007), "Energy lifecycle inputs in food systems: A comparison of local versus mainstream cases", *Journal of Environmental Policy Plan*, Vol. 9, No. 1.

Verghese, K., H. Lewis and L. Fitzpatrick (2012), *Packaging for Sustainability*, Springer, London.

Wallgren, C. and M. Höjer (2009), "Eating energy - identifying possibilities for reduced energy use in the future food supply system", *Energy Policy*, Vol. 37, Issue 12.

Wirsenius, S., F. Hedenus and K. Mohlin (2011), "Greenhouse Gas Taxes on Animal Food Products: Rationale, Tax Scheme and Climate Mitigation Effects", *Climatic Change*, Vol. 108, pp. 159-184.

Weber, C.L. and H.S. Matthews (2008), "Food-miles and the relative climate impacts of food choices in the United States", *Environmental Science and Technology*, Vol. 42, No. 10.

Worrell, E., J. Laitner, M. Ruth and H. Finman (2001), "Productivity Benefits of Industrial Energy Efficiency Measures", *Energy*, Vol. 28, No. 11.

Wreford, A., D. Moran and N. Adger (2010), *Climate Change and Agriculture: Impacts, Adaptation and Mitigation*, OECD Publishing, Paris, http://dx.doi.org/10.1787/9789264086876-en.

Annex 1A.1

Defining the terms: Methodological considerations

The challenge of measuring energy usage

Tracking trends in energy efficiency and comparing the performance of countries is made challenging by the lack of a single indicator to measure energy efficiency levels and changes. Instead, in the energy balance for a given production process, a variety of indicators may serve and support energy efficiency analysis (IEA, 2014b). Possible indicators include: primary energy per cultivation area or per tonne of agricultural product, or energy contained in agricultural products divided by the energy consumed (OFAG, 2015). Although the cultivation area is often used as a denominator, the link between this variable and energy consumption is not very strong. For example, in the European Union, energy use by agriculture per cultivation area in hectares was highest in the Netherlands, which is mainly due to intensive greenhouse farming. This form of funding accounted for 80% of total energy consumption by agriculture in 2010, but occupies only 0.5% of the total cultivation area in the country (EUROSTAT, 2012).

The quantitative assessment of energy flows in food systems is most often carried out following one of two approaches: the life cycle analysis (LCA) and the input-output (IO) accounting (Burney, 2001). These two approaches differ fundamentally in both the conception and the data inputs, and therefore it is not surprising that results often differ. From the variety of metrics it is clear that energy analysis has nevertheless not evolved into an exact science, and this explains the few comprehensive studies of any sector let alone one as complex as food. Challenges and ultimately subjective decisions lay in determining boundaries, aggregating different forms of energy, and defining energy credits for by-products.

LCA, as a product-focused methodology, takes into consideration all energy inputs along the full production (and disposal) chain, wherever these occur and as such it isolates the direct and indirect energy requirements of specific products. All the steps involved in creating a certain product are analysed, starting from raw material extraction and conversion, then manufacture and distribution, to the final use and/or consumption. LCA also includes re-use, recycling of materials, energy recovery and ultimate disposal. Interest in forms of LCA, particularly energy use and emissions, has recently increased as retailers attempt to develop consistent "energy foot printing" labels for their products.

LCA, however, needs detailed data on product "history" and is sensitive to the definition of the boundaries of the production system and to the methodology used for allocating the embodied energy among co-products or by-products. LCA remains challenging when applied to large economic sectors as apparently "similar" products can be enormously diverse in reality.

Numerous LCA studies have shown the cumulative energy intensity of food products (e.g. de Vries and de Boer, 2010; Carlsson-Kanyama et al., 2003; Heller and Keoleian, 2000).[1] However, when comparing across studies an important methodological caveat is the boundary of the analysis can often vary.

The I-O analysis is a tool that can be used to provide estimates of inputs (including energy) per unit of final product based on how various sectors of an economy are linked and exchange resources (including energy) and can provide very precise results down to a certain level of aggregation, taking into account direct and indirect contributions. Nevertheless, I-O needs to be complemented with exogenous data as far as process steps taking place outside the studied economic area are concerned. However, the accuracy of this approach is also dependent on the accuracy of input-output data, which are prone to become outdated. Moreover, specific sub-sectors of the economy may not always be disaggregated to an extent that makes for consistent product comparisons, (see, for example, Canning et al., 2010).

Note

1. See also www.agrilca.com..

Chapter 2

Energy use and opportunities for energy efficiency on-farm

Energy, as a production input, is an essential element affecting the profitability and competitiveness of the agricultural sector. In addition, agriculture might become an important potential source of renewable energy and thus provide significant economic opportunities for farmers and the rural economy, as well as improving the environment. This Chapter focuses on the issue of energy use and efficiency in agriculture. It presents some empirical evidence on the current situation and trends in energy consumption and energy efficiency for the agricultural sector. It also discusses the efficiency gain potential associated with different product categories (e.g. arable crops, horticulture, meats and dairy) and a number of options to improve energy efficiency on-farm are considered.

The statistical data for Israel are supplied by and under the responsibility of the relevant Israeli authorities. The use of such data by the OECD is without prejudice to the status of the Golan Heights, East Jerusalem and Israeli settlements in the West Bank under the terms of international law.

Current situation and recent trends

Different forms of energy are used for different purposes. Farmers mainly use energy from fossil resources – either directly, in vehicles and machinery (including tractors, dryers and heaters), using fuel or electricity – or indirectly, with the use of energy embodied in the production of agricultural machinery, fertilisers or pesticides, plastics and feed. However, the different production systems in various environments across countries vary substantially in their energy-use and energy-saving potential.

Modest improvements in efficiency of direct energy consumption in OECD agriculture

For the OECD area as a whole, direct energy use in agriculture is around 2% of total energy consumption over 2011-13 – roughly in line with its contribution to GDP (Figure 2.1). Turkey, Poland, Denmark and the Netherlands were the countries with the highest share, while the share was lower than 1% in Japan, Luxembourg, Switzerland and the United Kingdom.

As a result of its low figure, it can be argued that in most OECD countries changes in direct energy consumption in the agricultural sector are unlikely to have major implications for the overall supply and demand for energy in the country concerned. However, within the agricultural sector, changes in the supply and demand of energy can have significant implications for the profitability of a country's agriculture, the mix of output and management practices, as energy and energy-intensive inputs account for a significant share of total production costs in most farm activities.[1]

In France, in 2010, energy use, both direct and indirect, accounted for about 13% of variable costs on average, with some sectors (e.g. horticulture) being as high as 20% (ADEME, 2012). However, horticulture and granivore energy consumption is relatively small. In Italy, agriculture accounts for around 3% of national energy consumption. Total energy use constitutes an important part of farmers' costs and is strongly related to the type and size of farming (Fabiani et al., 2016). In the United States, on average 15% of total farm production expenses are energy-related. This number varies depending on the type of farming, the geographic location of the farm, and the type of products and processes used (Brown and Elliot, 2005).

For the overall economy, a clear decoupling of economic growth from energy consumption can be observed over 1990-2012, both for OECD and non-OECD countries (Figure 2.2). Concerning agriculture in the OECD area, direct energy use increased, while improvements in energy efficiency were variable. A clear upward trend in energy efficiency is noticeable from 2006, possibly due to the rise in energy prices and heightened concerns relating to climate change. By contrast, in the non-OECD area, energy use increased from 2000 onwards, while energy efficiency has increased relatively consistently since 1990. However, differences between countries in changes in energy use and efficiency remain high (Table A.2.1).

The global monitoring framework developed by the World Bank for the 2014-24 SE4ALL initiative showed that different sectors, countries and regions have exhibited different rates of energy efficiency improvement (Charpentier, 2016). Between 1990 and 2010, the agricultural sector reached its highest rate of improvement – at 2.2% per year – while industry and other sectors of the economy improved their energy intensity at a rate of only 1.4% per year. Since the mid-1980s, the energy intensities in high-GDP countries have declined, due to the introduction of energy efficient practices that have allowed a continued increment in crop yields. While the average energy intensity of agriculture in low GDP countries is usually lower than in high GDP countries, in more recent years, the increased use of fertilisers and mechanisation in the People's Republic of China (hereafter "China") and India, in particular, has led to rising levels of energy intensities. Taking these opposing trends into account, it was concluded that overall global energy intensities started to decline slightly after the 1980s, though the trend varied widely between countries.

In the United States, energy use by agriculture peaked in 1978. However, rapidly rising energy prices caused by oil-price shocks in the early 1980s forced farmers to become more energy efficient. Since 1978, total energy use by the agricultural sector has been falling. Despite this decrease, agricultural output has been increasing since the late 1970s. Energy efficiency, as measured by the ratio of energy use to agricultural output, has fallen by about 50% since 1978.

Because of the diversity of farming systems across the OECD countries, comprehensive cross-country comparisons of total energy use and efficiency in farm sectors in the OECD area are scarce. In the European Union, livestock and dairy products (with the exception of milk) incorporate a substantial amount of energy, while vegetables and bread are less energy-intensive per kilogramme of product (Monforti-Ferrario et al., 2015).

Figure 2.1. Between 2-3% of total energy is consumed on-farm

Share of agriculture in total energy consumption, 2011-13 (%)

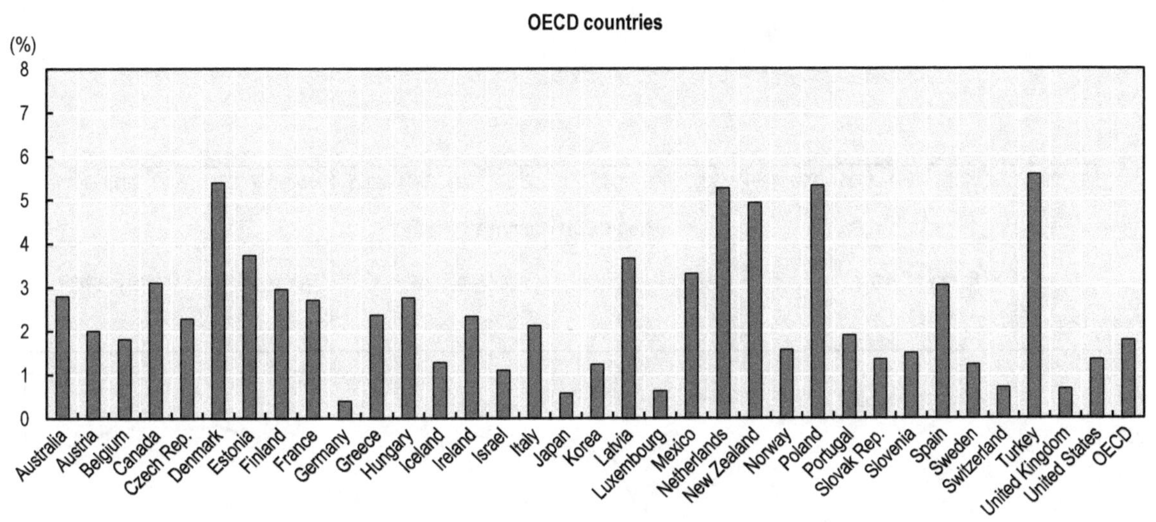

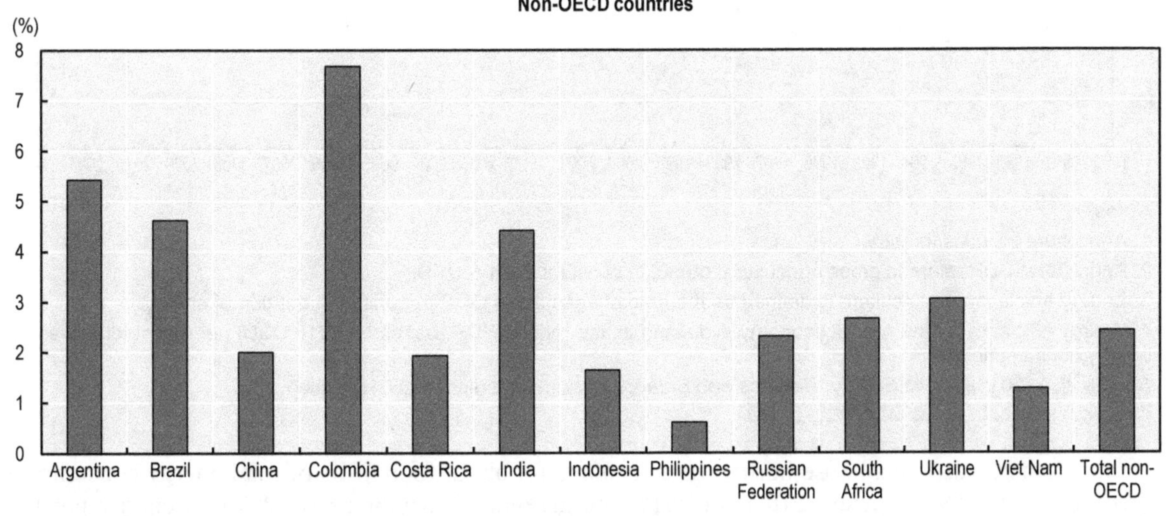

Notes:

1. Agriculture includes forestry.
2. No data for Chile.

Source: IEA (2016), *World Energy Statistics and Balances,* online data service 2016 edition, http://dotstat.oecd.org/?lang=en

Figure 2.2. Efficiency of direct energy consumption in agriculture is improving

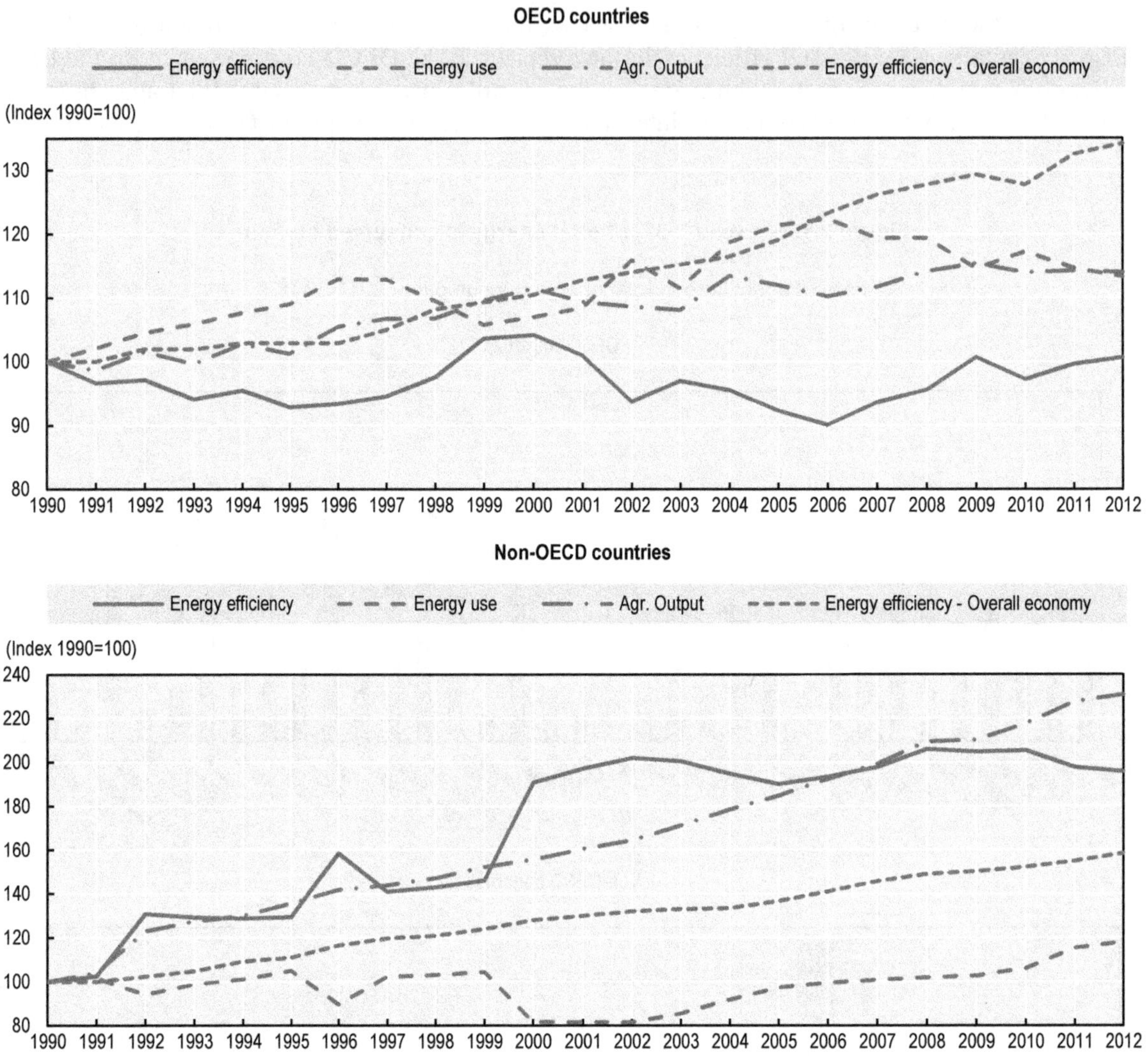

OECD countries

Energy efficiency — Energy use — Agr. Output — Energy efficiency - Overall economy

(Index 1990=100)

Non-OECD countries

Energy efficiency — Energy use — Agr. Output — Energy efficiency - Overall economy

(Index 1990=100)

Notes:

1. Agriculture includes forestry.
2. Agricultural GDP refers to gross agricultural output in constant 2004-6 USD.
3. Energy efficiency in agriculture is defined as the ratio of agricultural GDP per unit of direct use of energy.
4. Energy efficiency for the overall economy is defined as the ratio of GDP (based on 2010 USD) per unit of total final energy consumption.

Source: IEA (2016), *World Energy Statistics and Balances,* online data service 2016 edition, http://dotstat.oecd.org/?lang=en; FAOSTAT.

In the Netherlands, greenhouse horticulture accounts for about four-fifths of total energy consumption of the agro-food sector (NL Agency, 2012). In 2011, the greenhouse sector used 52% less energy per unit of production compared with 1990 (OECD, 2016b). This was partly achieved through the implementation of combined heat and power installations that both decreased the energy use per unit of production and generated income from electricity sales. The government has also played a role in the successful transition to more efficient energy use in the greenhouse sector. The ministry works together with the greenhouse horticulture sector in an innovation programme and provided some seed money in the past to stimulate energy development, and decreased the tax rate the sector pays on gas consumption, as it did for other energy-intensive sectors. In exchange the CO_2-sector system is regulated by law, by which a maximum CO_2-emisson is set. The energy transition agreement 2014-20 between the greenhouse sector and the Ministry of Economic Affairs is focused on absolute reduction of CO_2-emission by stimulating energy saving and the use of renewable energy. The ambition is a climate neutral greenhouse sector in 2050. The ministry and the greenhouse sector each provide 50% of funding for research activity on energy savings and geothermal energy. The CO_2 emission has declined by 30% in the period 2010-15.

In the United Kingdom, the main energy-using sectors reported by DEFRA (2008) were: arable crops, including potatoes and sugar beet (43%); protected horticulture, including mushroom production (28%); beef and sheep (14%); and poultry meat and eggs (8%). The dairy, field horticulture and pigs sectors each accounted for less than 3% of the total. Greenhouse horticulture, and the beef and sheep sectors use natural gas and gas oil, respectively, as their main energy sources.

Energy consumption and efficiency also vary in Switzerland, according to on the type of farm holding (OFAG, 2015). In 2013, the energy consumption of the agricultural sector in 2013 amounts to approximately 53 900 terajoules (TJ), of which about 30% is direct energy consumption. Most of the energy consumption is related to infrastructure such as buildings and farm machinery. Consumption of transport fuels accounts for the largest share of direct energy (about 39%), followed by other fuels (mazut and gas) (33%). Finally, electricity accounts for 22% of direct energy consumption and renewable energies 6%. The heating of the agricultural greenhouses required considerably more energy (3 900 TJ), oil and gas combined, than the heating of stables (1 500 TJ).

In 2013, the total energy consumption of Swiss agriculture increased by 6% compared to 1990. After a slight decline initially, the figures increased continuously between 1999 and 2007 and remained more or less stable since. The total direct energy consumption hardly changed during this period. There has been a slight increase in electricity and fuel consumption, but this is offset by an equivalent decline in the consumption of fuel oil and gas. Consequently, the evolution of total energy consumption is mainly due to changes in indirect energy. These include the decline in the use of mineral fertilisers in the 1990s. On the other hand, since the end of the same period, imports of animal feed have increased, leading to an increase in embodied energy. Energy efficiency has remained stable since 1990, with vegetable production being more efficient than animal production.

Understanding the sources of energy use in the agro-food chain is vital in determining how energy savings can most readily be made. The amount and type of energy used in agricultural operations also affect overall CO_2 emissions, and generally CO_2 levels increase with higher energy use in agriculture.[2] Agriculture is heavily dependent on energy from fossil fuels, because it requires energy inputs at all stages of production in both the direct use of energy for agricultural machinery, irrigation and harvesting, and also in indirect use for post-harvest operations, such as processing, storage and transportation of agricultural products to markets.

Over time, the type of energy used by the agricultural sector of the OECD area has changed, with the direct use of fossil fuels and coal declining, and consumption of electricity and renewables (biofuels and waste) increasing (Table A.2.2). On average, in 2011-13, 68% of direct energy consumed by the agricultural sector in the OECD area originated from fossil fuels; 16% from electricity; 9% from natural gas and 4% from renewable energy sources (Figure 2.3).[3]

Figure 2.3. Fossil fuels are the main source of on-farm direct energy consumption in the OECD area

(2011-13, %)

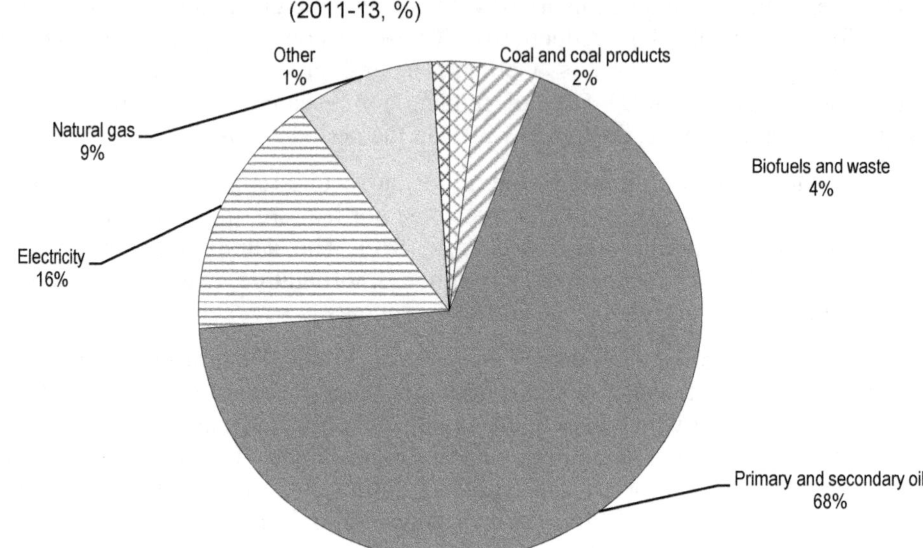

Notes:

1. 1990= 100.
2. Agriculture includes forestry.
3. "Primary and secondary" oil includes crude and refined oil products.

Source: IEA (2016), *World Energy Statistics and Balances,* online data service 2016 edition, http://dotstat.oecd.org/?lang=en.

Different types of energy are often required for different activities in food production

Energy use in agriculture varies across agricultural operations by crop or livestock type, size of operation, and geographic location. Energy use also varies over time, depending on weather conditions, changes in energy prices, and changes in total annual crop and livestock production.

Direct energy, mostly in terms of refined petroleum products, is used on farms for planting and harvesting, fertiliser and pesticide application and transportation. Direct energy is also used for land preparation, tillage, post-harvest storage of harvested crops and pre-processing activities (e.g. drying).

Fuel use for field operations, irrigation, drying and storage of crops are the most important direct energy-demanding activities used in arable cropping systems. Dairies require fuel for cooling milk, operating milking systems and supplying hot water for sanitation. Natural gas is commonly used to control greenhouse temperatures, and dairies rely heavily on electricity to operating milking systems.

In general, direct energy consumption of greenhouse horticulture and intensive livestock – and, to a lesser extent, dairying – is mainly related to fixed equipment that provides energy for heating and lighting. In contrast, energy for field equipment – such as fuel for mobile agriculture machinery (e.g. tractors), self-propelled chemical application vehicles and potato harvesters – is prevalent in the arable crops sector. Nevertheless, certain crops require the use of some fixed equipment. For example, cereals and oilseeds can have significant energy-use needs for drying, storage and handling (Table A.2.3).

Empirical evidence on indirect energy use and efficiency is limited

Existing estimates of direct energy consumption only partially reflect the actual amount of energy consumed, as several inputs are not fully allocated to the agriculture sector's energy statistics. A more complete picture of energy use, and of energy-saving opportunities is gained by looking beyond direct farm energy use, to consider both inputs and outputs. The production of agricultural inputs can be heavily energy-intensive, most notably for fertilisers and pesticides.

In general, indirect energy consumption is usually significantly larger than direct energy consumption in intensive agriculture systems, mainly due to fertiliser and pesticide use (Monforti-Ferrario et al., 2015; Pelletier et al., 2011; DEFRA, 2008). As an example, DEFRA estimated that in the United Kingdom, for the period 2003-07, the agriculture sector consumed 2.5 times more energy indirectly than directly (DEFRA, 2008).[4] In Switzerland, indirect energy use accounts for almost 70% of total energy consumption in agriculture (OFAG, 2015). In contrast, a study by Beckman, Borchers and Jones (2013) reports that, for the United States over the 2001-11 period, indirect energy use is more important than direct energy use in terms of expenditures, and less important in terms of physical energy units.[5] A big part of this is due to the fact that direct energy inputs use much more energy per unit of input (gallon, litre, kg) than indirect energy sources.

Fertiliser use is a classic example of high indirect energy use. In arable farming, for example, nitrogen is often the biggest single energy input. Application of mineral fertilisers requires energy, but the fertiliser industry is also a consumer of energy. In the United States, for example, inorganic fertilisers accounted for slightly more than half of indirect energy use on US farms in 2011, while pesticides accounted for slightly less than half of the indirect energy used (Beckman, Borchers and Jones, 2013).[6]

In the United States, agriculture accounts for a fifth of the energy used in the US food system. According to estimates, agricultural food production represents more than a fifth (22%) of the total energy consumed in the US food system, including the energy expended in the manufacturing of chemical fertilisers and pesticides, along with the consumption of fossil fuels used in farming (Canning et al., 2010; Heller and Keoleian, 2000). Additionally, most tractors in the United States use gasoline or diesel. Therefore, improvements in efficiency can help reduce the dependence of agriculture on oil.

For the European Union, the comprehensive study by Gołaszewski et al. (2012) – applying a LCA-like approach – has provided results for seven EU members (Denmark, Finland, Germany, Greece, the Netherlands, Poland and Portugal) and for the following sectors: crops (wheat, sugar beet, potatoes, cotton and sunflower); greenhouse production of tomatoes, cucumbers and sweet peppers; perennial crops (vines and olive trees); and livestock production (dairy cows, pigs and broilers). Key findings of the study include:

- The efficiency of energy use in agricultural production is specific to country and locale. Energy consumption varies substantially for all of the crops and countries considered, due to the existence of various cropping practices, different types of agricultural machinery, varying yields and dissimilar climates.
- For field crops, the main energy input is the use of fertilisers and diesel. Energy inputs for irrigation, drying and/or storage are frequently important, but this depends on geographical location, climate and the intensity of the production systems.
- Greenhouse vegetable production in central and northern-EU countries is characterised by a very intensive direct energy input and differs significantly from the production systems in southern EU countries. For the crops grown in the southern countries, little or even no energy input is needed when they are grown directly on soil; a higher energy input is needed only for hydroponic systems.
- The production of olives requires higher energy use in south western-EU countries than in south eastern-EU countries.
- For livestock production, the main energy input is associated with feed, although this indirect input varies between countries and sector. Milk production is the most energy-consuming subsector,

followed by pig and broiler production. In milk and broiler production, energy use for feed varies across countries studied, while in pig production, energy use for feed is similar, although the total energy input depends on the level of direct energy inputs.

- Across the six EU countries studied, the most energy-consuming subsectors are dairy cows, wheat, and pig production. The most energy-consuming subsectors are concentrated in different counties: in the Netherlands – dairy cows, pigs, and tomato and sweet pepper production; in Poland – dairy cows, wheat, pigs and potato production; in Finland – dairy cows and pigs; in Greece – wheat and cotton, and in Portugal – dairy cows, olive groves and broiler production.

Empowering agriculture: Energy efficiency options

Opportunities and constraints related to energy efficiency measures in agriculture

Energy needs vary enormously across different farming systems, as do the number of options that could be employed to reduce energy use and increase energy efficiency. These broadly include: technical progress, input substitution in agriculture and demand changes for agricultural commodities (Schneider and Smith, 2009).

Technical progress can be achieved with respect to the energy efficiency of all major inputs. Principal strategies could include: more efficient machinery; improved agro-chemicals management (e.g. precision cropping with site specific management of nutrients, pesticides, and water; as well as computer controlled livestock feeding); more efficient irrigation systems; plant and livestock genetic improvements; improved insulation; efficient light sources; and more efficient heaters. Bio-energy and bio-material strategies can also be employed. For example, a large spectrum of dedicated energy crops, plant residues, livestock manure, and by-products of agricultural commodity processing could be converted into energy or industrial material – cellulose into bio-fuels; establish improved crop varieties for the production of industrial oils and bio-polymers – thereby reducing the consumption of and dependency on fossil energy.

On-farm energy savings and energy efficiency improvements can also be achieved with existing technologies through *substitution of inputs*.[7] Input substitutions are driven by economic conditions – foremost, by the cost of energy. If the relative price for energy increases, the overall energy intensity at a given production level will fall. However, the resulting substitution effects can be complex because energy is contained in varying degrees in almost all agricultural inputs.

In the short-run, possible input substitution options entail: changes in irrigation; switching from conventional tillage practices to reduced or no-till; reducing fertiliser application rates; improving nutrient management practices that reduce nutrient losses; reducing crop protection intensities and level of mechanisation; the early retirement of fuel inefficient machinery; switching to crops that use less fertiliser and other energy-related inputs; grass cultivation as part of crop rotation or as cover and catch crops; biogas production on farms from crop residues and manure; and livestock management alternatives related to feeding, housing, and manure treatment. In the long term, farmers have more flexibility in reducing their energy use by acquiring more energy efficient equipment and making other changes to their farming operations.

Demand changes for agricultural commodities influence the total volume of production and, thus, the total amount of energy used in the food system. The demand is driven by market prices, policies and consumer preferences. These drivers can promote changes in human diets towards foods that are less energy-intensive and increased demand for renewable energy and products. The most important driver on the demand side is changes in the relative price of energy (Canning, et al., 2017). Higher energy prices increase the wedge between energy friendly and energy intensive commodities and thereby shift consumption towards the former.

How producers respond to energy price changes depends primarily on producers' expectations of whether the price change is only temporary, or permanent. If producers perceive an energy price change as temporary (e.g. lasting only for the current crop season), their response may be limited to some small-scale efforts to economise on fuel use, perhaps by switching to fuel-saving cultivation methods (such as minimum or no-till production), or by applying smaller volumes of fertilisers and pesticides per hectare. However, the ability of producers to implement such changes is greatly diminished once a crop has been planted and the production strategy is in place. As a result, unexpected within-season price changes can have unavoidable time-limited impacts on farm income. On the other hand, if the change in energy price is perceived as permanent, producers are more likely to adopt more important responses, such as adjusting the activity mix and production practices to take into account the new revenue-cost structure.

Box 2.1. Energy efficiency options may be country-specific – The case of greenhouses

Greenhouse production systems use a substantial amount of energy indicating large potential for energy efficiency. Most of these energy-saving measures target added insulation and heat recovery systems, which have both economic and environmental benefits – but significant financial investments are typically needed to achieve the most efficient greenhouse systems. Additionally, in most cases energy-saving innovations cannot be implemented without careful consideration of local factors.

For example, greenhouse production in central and northern EU countries is characterised by highly intensive direct energy input due to the substantial direct-energy consumption required for heating. In the Netherlands, for example, the best way to improve the energy efficiency and environmental impact of glasshouses would be to lower the heating requirement of the building. In southern Europe, where far less heating is generally required, measures such as improved water and nutrient management would have a greater impact.

Research that has been carried out in the Netherlands shows there is an optimal design for greenhouses – in terms of investment and energy use – that is based on local climatic and market conditions. In the case of unheated greenhouses, these factors are even more important. As a result, greenhouse builders have to invest in design methodologies and software for modelling the energy performance of greenhouses.

Source: Golaszewski et al. (2012), State of the art on energy efficiency in agriculture - Country data on energy consumption in different agro-production sectors in the European countries, www.agree.aua.gr/Files/Agree_State.pdf

Economic studies have attempted to measure year-to-year farmer responsiveness to changes in prices. In the aggregate, studies suggest that a 10% rise in fuel prices is associated with a decline of about 6% in fuel use (Sands et al., 2011; Schnepf, 2004). Fertiliser and pesticide use are also negatively related to changes in their prices. A 10% rise in prices induces a 7% decrease in fertiliser use and a 5% decline in pesticide use.

Box 2.2. Significant energy savings in the US farming sector

- Producers with animal feeding operations can save up to USD 250 million nationwide each year by regularly maintaining their ventilation and heating systems and using more energy-efficient fixtures and equipment for animal housing.
- Converting irrigation systems from medium- or high-pressure to low pressure could cut energy costs by up to USD 100 million annually.
- Improving water efficiency by just 10% could reduce diesel consumption by 102 million litres and save farmers and ranchers USD 55 million annually.
- Doubling the amount of no-till acreage (from 25 million hectares to 50 million hectares) could save farmers and ranchers an additional 821 million litres of diesel fuel per year, valued at about USD 500 million annually.
- Doubling the application of manure-based nitrogen to replace fertiliser produced from natural gas could save USD 825 million and 2 831 billion of litres of natural gas annually.
- Reducing application overlap on 101 million hectares of cropland could save up to USD 825 million in fertiliser and pesticide costs annually, which would have knock-on effects on energy usage.

Source: Farm energy efficiency, www.nrcs.usda.gov/wps/portal/nrcs/main/national/energy/

Empirical evidence from the United States shows that farmers adapted to higher energy and fertiliser prices over the 2001-12 period by shifting to more energy-efficient production practices (Beckman, Borchers and Jones, 2013; Harris et al., 2008). Lower energy use was achieved through such measures as employing machinery less intensively and servicing engines more frequently, while lower consumption of fertiliser was achieved through the greater use of soil testing, changes in plant populations and the adoption of precision application methods. Farmers with the highest energy and fertiliser costs and the lowest net incomes were the most active in adopting measures to reduce input usage, suggesting that when changes in input costs become significant for farm profitability, farmers pay particular attention to addressing these.

Enhancing energy efficiency in the production of fertilisers

The bulk of energy use associated with fertilisers is not consumed directly on the farm, but indirectly, during its production, packaging and transportation to the site (Gellings and Parmenter, 2004). Other energy-consuming inputs include animal feed (largely due to the energy requirement of the palletisation process), hybrid seeds and water.

Globally, fertiliser production consumes approximately 1.2% of the world's total energy each year (IFA, 2016). Natural gas is used both as an energy source and as a feedstock for fertiliser production. Indeed, the majority of fertiliser energy-use is due to the use of natural gas as a feedstock for the production of ammonia for nitrogen fertilisers. Ammonia production is a highly energy-intensive process, requiring about 1 090-1 250 m^3 of natural gas to produce 1 metric tonne of anhydrous ammonia. Natural gas is also used as a fuel to generate heat for the production of other types of fertilisers.

Ammonia production accounts for the bulk (approximately 90%) of the fertiliser industry's total energy consumption. Feedstock for ammonia production continues to evolve in favour of natural gas and coal. In 2016, natural gas accounted for 70% of total ammonia production, while coal contributed 27%, and other feedstocks (naphtha, fuel oil and heavy residues, refinery off-gases and condensates, petroleum coke, hydrogen from water electrolysis) for 3%.

Energy-efficiency performance has continually improved since the mid-2000s (IFA, 2015).[8] In addition to some of the cross-cutting measures mentioned in Chapter 4, a number of measures are reportedly being taken by fertiliser manufacturers to increase both the process and feedstock energy efficiency of fertiliser production. The improvement of boiler efficiency, the reduction of steam loss and the recuperation of heat can result in significant energy savings and improved production, for example. Indeed, some of the decline in the indirect use of energy by farms as a whole in the United States between 2001 and 2011 is considered to be due to fertiliser manufacturers becoming more efficient in the use of natural-gas feedstocks and energy in the production of fertiliser (Beckman, Borchers and Jones, 2013).

The Haber-Bosch process enabled efficient production of ammonia, but the process is energy-intensive and entails a substantial carbon footprint

The limitation of plant available nitrogen for crop uptake has always been a central concern in agricultural systems. The Haber-Bosch process that is now used to produce nitrogen fertilisers allows the transformation of atmospheric nitrogen gas into ammonia under high pressure and temperature using natural gas (or coal) as a source of hydrogen and energy. Ammonia can also be produced without fossil fuels, in which case, hydrogen is obtained from water through electrolysis. The process, however, is not widely used due to its very high electricity requirements which make it very expensive, but it could develop in the coming decades. The large impact of the Haber-Bosch process on the increased productivity of agriculture and global food production over the last century is well recognised. Currently, it is estimated that 48% of the world population is fed through this process (Erisman et al., 2008). While discovery of the Haber-Bosch process in 1913 was a significant leap forward, the process is energy-intensive and releases large amounts of GHGs. Globally, ammonia fertiliser production represents almost 1.2% of total final consumption of energy and contributes nearly 1% of GHGs. The production of 1 ton of ammonia releases approximately 1.5 tons of CO_2. Decarbonisation of ammonia fertiliser production could be achieved through increasing the use of electrochemical processes and renewable energy sources, and by reducing the use of ammonia fertiliser itself, with yields being maintained through improved farming practices and management.

Saving energy in fertiliser production

In 2006, a large major US agro-food company, J.R. Simplot, invested approximately USD 180 000 in measures to reduce the energy used by the steam system in its Idaho fertiliser manufacturing plant. The operation of its boiler was optimised by reducing steam venting, improving condensate recovery, and repairing steam traps and leaks. As a result, the plant achieved energy savings of more than 75 000 MBtu (Mega-BTUs); a reduction in annual water consumption of more than 11.4 million litres; and total annual cost savings of USD 335 000, securing energy savings greater than the installation costs enabling simple payback within approximately 6.5 months (USDA, 2008).

A more targeted use of agro-chemicals and manure by farmers

Given the large volumes of energy used as a feedstock for the production of fertiliser, its optimised application by farmers is one of the key means to reduce energy use within the agro-food chain. Although fertilisation is often required to increase crop production, multiple studies have demonstrated that farmers often practise inefficient over-application. There is potential for farmers to take actions that generate "triple wins" in the form of energy efficiency, cost and emissions savings (e.g. synchronisation of applications to the growth needs of plants and improved nitrogen-cycling, using manures).

In addition to the energy-efficiency gains resulting from the optimal use of fertiliser and other energy-intensive inputs such as pesticides, improved fertiliser – and, indeed, pesticide and manure – management also generate important environmental benefits in the form of reduced air and water pollution. The reduction of water pollution, in particular, generates further energy efficiency benefits by reducing the need for energy-consuming water treatment. In 2009, for example, it was estimated that farmers in the United States were applying approximately 20-30% more nitrogen fertiliser than necessary, and that the reduction of nitrogen fertiliser use would additionally reduce downstream water pollution by 20-30% (NRDC, 2009).

Farmers have been exploring various means to enhance energy efficiency through: nutrient management, including the planting of local vegetation species, where possible; efficient use of biological nitrogen fixation by growing, for example, nitrogen-fixing legume crops, such as clovers, in pasture, or lupins as green crops; the uniform application of broadcast fertiliser; certain crop rotations (a low nitrogen-requiring crop, such as soybeans, may require little or no fertiliser if planted following a heavily fertilised maize crop, for example); timing of fertiliser application to coincide with the crop's period of greatest uptake; liming; the incorporation of manure directly after application (Extension, 2012);[9] and the use of compost or liquid slurry resulting from the anaerobic digestion of manure, for example. In certain cases, the slurry produced by anaerobic digesters has been found to act as an efficient replacement for artificial fertilisers, enabling greater uptake of nitrogen by crops (NFU, 2010).

Finally, chemical fertiliser and pesticide companies are both investing heavily in agricultural extension programmes to disseminate advice to farmers on the "correct" use of fertilisers and pesticides. Examples of industry-sponsored extension programmes in fertiliser and pesticide use are the product and nutrient stewardship programmes operated by the relevant industries, such as the international fertiliser industry (IFA)'s 4R Nutrient Stewardship Programme. This programme advocates the right nutrient use at the right rate, right time and right place (www.nutrientstewardship.org; CNW, 2014).

A simple reduction in fertiliser application is another possible energy-efficiency measure, which can be effective to a certain extent (Meyer-Aurich et al., 2012). In Denmark, farmers are obliged to apply nitrogen fertiliser at a rate of 90% of the economic optimum.

Promoting less energy-intensive soil tillage practices

Reducing tillage is a well-known and effective measure to reduce energy input in arable systems. Associated benefits typically include lower costs for the farmer and lower GHG emissions. Furthermore, a potential carbon sequestration effect, through storage of soil organic matter due to reduced tillage, may further mitigate the net GHG emissions from agriculture (OECD, 2016a).

In rain-fed agricultural systems, machinery use – mainly for soil cultivation and harvesting processes – is usually the second-highest user of energy. One means by which the energy use of machinery can be reduced is through the use of no- or reduced-tillage systems. Fuel use in no-till operations is invariably less than that used with conventional ploughing systems, although the degree of fuel use reduction will depend heavily on the soil type, the depth of ploughing, as well as the number and type of secondary cultivations (OECD, 2016a). No-tillage can also have the advantage of enabling reduced GHG emissions and improving the water permeability of the soil (Rusus, 2014). In some contexts, "no-till" methods can reduce fuel consumption for cultivation by 60-70% and total machinery investment can be reduced by 50%. It can also reduce losses of soil carbon (FAO, 2011). These cost reductions usually result in immediate higher profits when adopting a no-till system, even in those cases where yields, for example, do not respond positively in the first few years.

In addition, no-tillage practices allow large farms to use technological advances, such as controlled traffic farming and GPS-based precision farming, that lead to higher levels of efficiency of energy and input use (OECD, 2016a). These efficiencies have led some countries to implement policy initiatives such as the carbon credit scheme for offset markets from conservation tillage that has been operating in the Canadian province of Alberta for several years. The scheme integrates controlled traffic farming and GPS-based precision farming.

Reduced tillage has a number of disadvantages, however. Of these, high herbicide use is the most important, as it is often used to minimise the growth of weeds that would otherwise be controlled (at least in large-scale farming) by ploughing and other more intensive tillage methods. Herbicide production itself entails energy use. Nevertheless, it is argued that on average, burn-off herbicides require less energy than tillage (Smith et al., 2008). Second, depending upon soil and climate conditions, reduced tillage may also lead to lower crop yields. In addition, increased reliance on mulch inputs – which have competing uses as livestock feeds – has been observed in some cases. The resulting energy balance is determined by what feeds are being displaced.

Although adoption of no-tillage techniques is rising rapidly in several countries, with the exception of a few countries (e.g. the United States, Canada, Australia, Brazil, Argentina, Paraguay and Uruguay), it has not been "mainstreamed" by farmers or policy makers, and the total arable area under conservation agriculture worldwide remains relatively small (about 9%) (OECD, 2016a). The main factors hindering greater adoption, as cited in the literature, include: i) insufficient knowledge about the practice; ii) farmer attitudes and aspirations; iii) lack of appropriate machines; iv) lack of suitable herbicides to facilitate weed management; v) the high opportunity cost of crop residues for feed; vi) lack of herbicide-tolerant crop varieties for some crops and climates; and vii) inappropriate policies (e.g. commodity-based support in some OECD countries).

A more detailed discussion on the effects of reduced or no-till farm management practice on resource productivity and efficiency is provided in OECD (2016a), *Farm Management Practices to Foster Green Growth* (Chapter 2, "The role of soil and water conservation in the transition to green growth").

Using precision farming techniques

A further way to improve energy efficiency in agriculture is via the use of precision technologies which calculate the optimal quantity, timing and location for the application of inputs and thereby create economies related to both fuel and fertiliser use. Precision farming is a relatively new management practice which has been made possible by the development of information technology and remote sensing. It is a whole-farm management approach with the objective of optimising returns on inputs, while improving agriculture's environmental footprint. Different technologies, however, have different potentials, which need to be addressed specifically with regard to the effects on energy-use efficiency and environmental effects.

The precision agriculture management approach currently relies almost entirely on the private sector, which offers services, devices and products to the farmers. Public-sector involvement is generally very limited, notwithstanding the growing policy interest in the role of innovations for increasing productivity sustainability. An example of a recent public initiative aimed at the "mainstreaming of precision farming" is the creation of a focus group in the European Union under the European Innovation Partnership on Agricultural Productivity and Sustainability. The initial priorities of the group are to look at data capture and processing, but it is envisaged that the process will be expanded to encompass evidence-based benchmarking of precision-agriculture performance and impact evaluation.

A wide range of technologies is available, but the most widely adopted precision farming technologies, e.g. GPS guidance, are knowledge-intensive. Available data on adoption rates are fragmented and often dated because countries do not regularly collect data on the use of precision agriculture, while manufacturers and dealers in precision agriculture rarely reveal their sales data.

Adoption of precision-agriculture technologies is limited to only a few countries and sectors. Precision agriculture is most advanced amongst arable farmers – particularly those with large farm sizes – in the main arable-crop growing areas of Europe, the United States and Australia, who have well-developed business models that maximise profitability. Knowledge and technical gaps, high start-up costs with a risk for insufficient return on investment, as well as structural (e.g. small farm size) and institutional constraints are key obstacles to the adoption of precision agriculture by farmers (OECD, 2016a).

A more detailed discussion on the effects of precision agriculture on resource productivity and efficiency is provided in OECD (2016a), *Farm Management Practices to Foster Green Growth* (Chapter 6, "Is precision agriculture the start of a new revolution?").

Exploring the energy-saving potentials of organic farming

In several respects, energy use – for transporting grain by road or operating heavy machinery, for example – is not considered to differ significantly between conventional and organic farms. Nevertheless, a number of studies claim that organic systems are indeed more energy-efficient, largely due to the energy required to manufacture, ship, and apply pesticides and nitrogen-based fertilisers (OECD, 2016a). Organic agriculture can also – depending on soil type – enable greater water conservation (and therefore reduced energy consumption) by increasing the amount of organic matter, which reduces runoff while increasing crop yields (Pimentel, 2006; 2009).

On average, organic farming appears to be more energy-efficient than conventional farming, both in terms of energy per area and per unit of output. The impact of organic agriculture on energy use can be analysed on the basis of different functional units such as "area", or the weight of output from the farming system as a reference.

Smith et al. (2015) review of 50 studies found that overall most organic farming systems are more energy efficient than their conventional counterparts, although there are some notable exceptions. Organic farming performs better than conventional for nearly all crop types when energy use is expressed on a unit of area basis, but results are more variable per unit of product due to the lower yield for most organic crops. For livestock, ruminant production systems tend to be more energy efficient under organic management due to the production of forage in grass–clover leys. Conversely, organic poultry tend to perform worse in terms of energy use as a result of higher feed conversion ratios and mortality rates compared to conventional fully housed or free-range systems. With regard to energy sources, there is some evidence that organic farms use more renewable energy and have less of an impact on natural ecosystems. Human energy requirements on organic farms are also higher as a result of greater system diversity and manual weed control.

Lampkin (2007) identified that most product- and area-related energy use assessments of organic farming to date show lower energy use per hectare. Lower energy use (direct and indirect) for organic farming can also be found in France, particularly for arable crops and dairy, but energy usage remains higher for horticulture (Bellon and Penvern, 2014). Gomiero et al. (2011) describe energy use in different agricultural settings and conclude that organic agriculture has higher energy efficiency, but, on average, exhibits lower yields and hence reduced productivity. Tuomisto's et al. (2012) meta-analysis reports that energy use per unit of output is, on average, 21% lower than with conventional farming, although the results depend largely on the productivity levels of the systems examined, as well as on the types of production.

A similar result was found in the case study of the AGREE project in the Netherlands: energy consumption per unit of milk production under organic farming was 13% lower, than conventional farming while yields fell by up to 6.5% (from 8 500 L/LU/year to 7 950 L/LU/year) (Golaszewski, et al., 2012).

Schader (2009) examined the differences in energy use per ha between organic and conventional farm types in Switzerland, based on a representative farm sample. As well as pig and poultry farms, conventional mixed farms have the highest total level of energy use (60 GJ/ha), while average energy use, expressed as a sum of all energy-use components in dairy, suckler cow, other grassland, arable and speciality crop farms, ranges from 20 to 30 GJ/ha. The energy use of organic farms is about one-third lower (10-20 GJ/ha), except on mixed farms, where average energy use is approximately 50% less than on conventional farms. The author attributes the lower quantities of purchased feedstuffs (particularly concentrates), to lower stocking densities, the ban on mineral nitrogen fertilisers, and the absence of highly intensified specialised pig and poultry farms.

Improving feed-use efficiency offers scope for reducing energy use

The production of livestock protein is a further source of indirect energy consumption in the agricultural sector. Energy use per unit of caloric output in intensive livestock production is significantly greater than for agricultural crops, due to inefficiencies in the biological conversion of feed into fat or protein (Pelletier, 2011). Energy requirements for animal feed are the major determinant of the energy intensity of livestock production. In the United Kingdom, for example, animal feed accounts, on average, for 75% of total energy embedded in livestock (Woods et al., 2010).

Means to improve feed use efficiency – sourcing energy-efficient feed inputs and improving feed conversion ratios – offer scope for providing economic gains and reducing energy use. Improving feed conversion efficiency also decreases methane emissions and avoids feed crop wastage because methane production also decreases (Monforti-Ferrario et al., 2015).

There can, unfortunately, also be trade-offs between GHG emissions and energy efficiency. For example, while grain is considered by some to be a more energy-intensive feed than grass, grass-fed cattle produce more methane. Nevertheless, pasture-based systems are considered by some to have lower GHG emissions overall than conventional feed systems, due to their carbon sequestration potential (Profita, 2012).

Also the intensification of feed ratios for ruminants has its drawbacks and the limits of energy-efficient feeding strategies in ruminant production systems should be taken into consideration. For example, for the pork and poultry production systems – which are the most industrialised agricultural production systems – the economically most cost-efficient feeding strategy may not be effective from an energy-efficiency and GHG emission point of view. Such trade-offs relate to different sub-systems and sub-processes (manure system, heating system and feeding system), suggesting that measures on energy efficiency should also be cross-checked for compliance with consumer demand for animal welfare.

Finally, the determinants of the energy use of each species of livestock are influenced by breeding stock fecundity, the efficiency with which it converts feed into fat or protein, and in the case of beef cattle, the life-span of the animal, with energy usage greater for animals with shorter life spans. It also decreases, depending on whether manure is used for heat generation (Pelletier et al., 2010).

Poultry systems appear to offer the highest feed-use efficiencies, although some studies suggest that these can be lessened if animal by-products are used in the feed. This practice is used in the US but not in the European Union, where it is prohibited. In terms of manure use, Nguyen et al. (2010a) report, that in the European Union, manure use for energy generation in pig farming can potentially offset 57% of the supply chain energy use. However, the highest savings potential are reported when manure use is combined with improvements in feed use and manure management.[10] This could save up to 61% of fossil energy. For ruminants, Nguyen et al. (2010b), using a life cycle approach, showed that the energy requirements for calf production in suckler herds is considerably higher than in dairy herds.

In dairy, both replacement rates and milk yields are important determinants of energy use. The energy/yield relationship appears to follow an arc that first improves then deteriorates past a yield threshold of around 8 000 kg/milk/cow/year. In dairying, energy demand also increases with higher replacement rates. Capareda et al. (2010) surveyed 14 US dairies and found that energy use varied from as low as 464 kWh per year per animal ($kWhyr^{1}hd^{-1}$) in pasture-based systems, to as high as 1 637 $kWhyr^{1}hd^{-1}$ in a hybrid facility. Energy potentially produced from manure would exceed daily requirements in all cases, although there is some contradictory evidence that this may not hold for confined systems.

Cost-effectiveness savings in direct energy use in primary agriculture

There are few consistent empirical analyses of energy efficiency and the benefits of energy saving. One of the more revealing pieces of analysis has been conducted by Gołaszewski et al. (2012) for seven EU countries. It was found that the implementation of the energy-saving measures[11] can reduce both direct and indirect energy inputs and the overwhelming majority of these measures (443 out of 481) were assessed in the range from moderate to high in terms of their importance for energy saving. The implementation of the energy-saving measures in agricultural practice is achievable at present (464 out of 481 measures) but will require some advanced research (389 out of 481). In the highly industrialised production of pigs and broilers, there are many energy saving measures which may be implemented with technologies which are present on the market such as improved heat insulation, more efficient ventilation, lighting and cooling systems, as well advanced control of the interior climate.

The estimated categories of investment costs related to implementation of energy saving measures vary greatly between subsectors. One-third of the total number of the measures can be implemented at a cost under EUR 1 000, and one-third incurs in costs in the range of EUR 1 000 to EUR 25 000. The highest investment costs would be associated with saving energy and improving energy efficiency in greenhouse and livestock production. They are associated with improved heat insulation, more efficient ventilation, lighting and cooling systems, as well advanced control of the interior climate. Many technologies (e.g. heat exchangers and variable speed pumps) are “win-win” methods that can be implemented immediately and can achieve payback times of five years of less (DEFRA, 2013).

In arable systems major energy efficiency measures are associated with the use of direct energy through fuel usage and indirect energy use through nitrogen fertiliser usage. The analysis of different case studies across Europe showed energy saving potentials from 1% to 43% of the total energy use.

The importance of energy saving activities may be country-specific. In the southern EU countries greater importance will be attributed to the energy saving measures associated with irrigation of cultivated crops, while in the central and north-eastern countries, similar effort would be attached to energy saving measures associated with energy effective drying techniques.

The study conducted for DEFRA (2010) attempted to assess the cost-effectiveness of energy reduction methods in English agriculture for the years 2020 and 2030. The analysis aimed to locate energy reduction measures on the pre-existing agricultural GHG marginal abatement cost curve (MACC) which shows the cost effectiveness of alternative measures to reduce emissions (CO_2e) GBP/tonne. In essence, what the analysis shows is the extent of cost-saving achievable by the application of specific measures over and above their current (or baseline) level of use across the sector.

A revealing element of the pre-existing MACC was the demonstration of win-win measures (i.e. measures that could simultaneously reduce emissions while saving cost to the farm enterprise). A classic example is the over-application of nitrogen fertiliser beyond plant growth needs. Analysis of direct energy uses reveals that large cost (and emissions) savings can be achieved from a variety of measures (see Table A2.4).

The limited number of options for field operations may seem surprising since these are a large source of the emissions arising from farm energy use. However, most of this energy is consumed by tractors and there are only a limited number of options that can be adopted to improve tractor efficiency. It is also worth noting that many of the initial savings are estimated to sit under the headings of "energy management and maintenance"; these improvements are actually attributable to better operation of tractors and implements.

Note that this analysis did not include radical approaches such as precision farming, controlled traffic farming and minimum tillage systems. The problem with these alternatives (perhaps with the exception of precision farming) is that they are likely to meet with some significant resistance to uptake as they will require a shift change in the farming systems, in particular equipment requirements, that are currently in use.

Notes

1. Data from DEFRA and the USDA suggest that energy costs account for about 17% of farm production expenditure in the United Kingdom and the United States (for the latter, about 6.3% were direct expenses and 11.5% indirect). Moreover, maize, sorghum and rice farmers in the US allocated over 30% of total production expenditures on energy inputs in 2011 (Beckman, Borchers and Jones, 2013).

2. Some fuels have a higher carbon content than others, resulting in higher CO_2 emissions per btu (British thermal unit) used. However, some fuel/engine applications are more energy-efficient than others and require less fuel to perform similar operations. For example, diesel fuel has a higher btu content than gasoline on a volumetric basis, but diesel engines have a higher performance rating compared to gasoline engines.

3. In the United Kingdom, the energy use of agriculture in 2010 was estimated to be equivalent to around 4% of the energy used to generate electricity in the country (DEFRA, 2010). Gas oil (agricultural diesel) estimated to be 39% of total energy use is the most widely used fuel, followed by natural gas (17%), electricity (16%) and static oils (15%). Other energy requirements were supplied by coal and liquid propane gas.

4. More precisely, in 2007, the most recent year for which such an analysis is available, direct energy consumption in agriculture amounted to 839 ktoe, while 1 053 ktoe and 321 ktoe were consumed (respectively) for fertiliser and pesticide production, 503 ktoe for animal feed and 373 ktoe for tractors and other agricultural machinery.

5. In 2011, direct energy use in physical units (btu - British thermal units) accounted for 63% of agricultural energy consumption, compared with 37% for indirect use.

6. The study considers other inputs, such as electricity and natural gas, to be direct energy inputs.

7. Note that there is a fundamental difference between the economic interpretation of technical progress and input substitution. While the former shifts a production possibility frontier for a given input endowment outward, the latter involves movements along a given frontier.

8. According to IFA, the current average net energy efficiency for the 66 ammonia plants that participated in 2015 is 36 GJ/t NH_3, which is equivalent to a 4% improvement on the 2004 base year (IFA, 2015).

9. In Finland, intensive use of legumes could reduce use of fertilizer N by 60% and fossil energy use by about 3 700 TJ year-1 as compared to the current situation (Känkänen, 2015).

10. The authors used a consequential life cycle approach (LCA), which takes into account all processes that are affected by a change in the production of a product.

11. Energy efficiency measures refer to the reduction of main energy inputs, including fertilisers, pesticides and feed; transportation fuels for tractors and other machinery; fuel use for heating, cooling and ventilation in farm buildings and facilities; electricity use for pumping, lighting; and energy embodied in buildings and equipment (Gołaszewski et al., 2012).

Bibliography

Agence de l'Environnement et la Maîtrise de l'Énergie (ADEME) (2012), « Analyse économique de la dépendance de l'agriculture à l'énergie », www.ademe.fr/analyse-economique-dependance-lagriculture-a-lenergie-evaluation-analyse-retrospective-depuis-1990-scenarios-devolution-a-2020.

Alluvione, F., B. Moretti, D. Sacco and C. Grignani (2011), "EUE (energy use efficiency) of cropping systems for a sustainable agriculture", *Energy*, Vol. 36, No. 7.

Astier, M., Y. Merlín-Uribe, L. Villamil-Echeverri, A. Garciarreal, M. Gavito and O. Masera (2014), "Energy balance and greenhouse gas emissions in organic and conventional avocado orchards in Mexico", *Ecological Indicators*, Vol. 43.

Beckman, J., A. Borchers and C. Jones (2013), *Agriculture's Supply and Demand for Energy and Energy Products*, EIB-112, U.S. Department of Agriculture, Economic Research Service, May.

Bellon, S. and S. Penvern (eds.) (2014), *Organic Farming, Prototype for Sustainable Agricultures*, Springer.

Brown, E. and R. Elliot (2005), "On farm energy use characterizations", *American Council for an Energy-Efficient Economy (ACEEE)*, Report No. IE052.

Canning, P., S. Rehkamp, A. Waters and H. Etemadnia (2017), The Role of Fossil Fuels in the U.S. Food System and the American Diet, Economic Research Service, United States Department of Agriculture, ERS Report No. 224.

Canning P, A. Charles, S. Huang, R. Polenske and A. Waters (2010), *Energy Use in the U.S. Food System* Economic Research Service, United States Department of Agriculture, ERS Report No. 94.

Capareda, S. Mukhtar, S. C. Engler and L. Goodrich (2010), "Energy usage survey of dairies in the southwestern United States", *Applied Engineering in Agriculture*, Vol. 26, No. 4.

Charpentier, O. (2016), *Efficient Energy Use in Agri-food Chains*, Inter-American Institute for Co-operation on Agriculture (IICA), www.iica.int/sites/default/files/publications/files/2016/B3984i.pdf.

Canada Newswire's News (CNW) (2014), "Fertilizer industry and energy company promote greenhouse gas emissions reduction and agricultural sustainability for Farming 4R Land" (8 April), www.newswire.ca/news-releases/fertilizer-industry-and-energy-company-promote-greenhouse-gas-emissions-reduction-and-agricultural-sustainability-for-farming-4r-land-514052541.html, accessed September 2015.

Department for Environment, Food and Rural Affairs (DEFRA) (2013), *Energy Dependency and Food Chain Security*, Report FO0415, London, United Kingdom, http://randd.defra.gov.uk/Default.aspx?Menu=Menu&Module=More&Location=None&Completed=0&ProjectID=16433.

DEFRA (2010), *Energy Marginal Abatement Cost Curve for English Agricultural Sector*, final report, http://randd.defra.gov.uk/Default.aspx?Menu=Menu&Module=More&Location=None&ProjectID=17631&FromSearch=Y&Publisher=1&SearchText=EC0103&SortString=ProjectCode&SortOrder=Asc&Paging=10%23Description.

DEFRA (2008), *Comparative Life Cycle Assessment of Food Commodities Procured for UK Consumption through a Diversity of Supply Chains*, research undertaken for Defra by AEA, ADAS, Ed Moorhouse, Paul Watkiss Associates, AHDBM, Marintek, Defra project FO0103, 2008.

Dodder, R., A. Elobeid, T. Johnson, P. Kaplan, L. Kurkalova, S. Secchi and S. Tokgoz (2011), *Environmental Impacts of Emerging Biomass Feedstock Markets: Energy, Agriculture, and the Farmer*, CARD Working Paper 11-WP 526.

Erisman, J., M. Sutton, J. Galloway, Z. Klimont and W. Winiwarter (2008), "How a century of ammonia synthesis changed the world", *Nature Geoscience*, Vol. 1.

Extension (2012), *Energy-Efficient Use of Fertilizer and Other Nutrients in Agriculture*, http://articles.extension.org/pages/62014/energy-efficient-use-of-fertilizer-and-other-nutrients-in-agriculture.

Fabiani, S., O. Cimino, P. Nino, S. Vanino, F. Lupia and F. Altobeli (2016), "Energy Impact Matrix: using Italian FADN to estimate energy costs impact at farm level", paper presented at the 7th International Conference on Agricultural Statistics, 26-28 October, Rome.

Fertilizers Europe (2014), "Energy efficiency and greenhouse gas emissions in European nitrogen fertilizer production and use", http://www.fertilizerseurope.com/fileadmin/user_upload/publications/agriculture_publications/Energy_Efficiency__V9.pdf.

Food and Agriculture Organization of the United Nations (FAO) (2011), *Energy Smart Food for People and Climate – Issue Paper*, Rome.

Gelfand, I., S. Snapp and G. Roberson (2010), "Energy efficiency of conventional, organic and alternative cropping systems for food and fuel at a site in the US Midwest", *Environmental Science & Technology*, Vol. 44, No. 10.

Gellings C. and K. Parmenter (2004), "Energy Efficiency in Fertilizer Production and Use", *Efficient Use and Conservation of Energy*, in Encyclopedia of Life Support Systems (EOLSS), Developed under the Auspices of the UNESCO, Eolss Publishers, Oxford, United Kingdom, http://www.eolss.net/ebooks/sample%20chapters/c08/e3-18-04-03.pdf.

Gołaszewski, J., C. de Visser, Z. Brodzinski, R. Myhan, E. Olba-Ziety, M. Stolarski, G. Papadakis (2012), "State of the art on energy efficiency in agriculture - Country data on energy consumption in different agro-production sectors in the European countries", Project founded by the FP7 Programme of the EU with the Grant Agreement Number 289139, AGREE Project Deliverable 2.1., www.agree.aua.gr/.partnersarea/index-3.html.

Gomiero, T., D. Pimentel and M.G. Paoletti (2011), "Environmental impact of different agricultural management practices: Conventional vs. organic agriculture", Critical Reviews in Plant Sciences, Vol. 30, Nos. 1-2.

Harris, J., K. Erickson, J. Dillard, M. Morehart, R. Strickland, R. Gibbs, M. Ahearn, T. Covey, F. Bagi, D. Brown, C. McGath, S. Vogel, B. Williams, and J. Johnson (2008), *Agricultural Income and Finance Outlook*, AIS-86, Economic Research Service, U.S. Department of Agriculture.

Heller, M. and G. Keoleian (2000), "Life Cycle-Based Sustainability Indicators for Assessment of the U.S.", *Food System*, The Center for Sustainable Systems, Michigan, United States.

International Fertilizer Industry Association (IFA) (2016), *Fertilizer Market Outlook report: Fertilizers and Raw Materials Global Supply 2016-2020,* www.fertilizer.org/

IFA (2015), *Energy Efficiency and CO_2 Emissions in Ammonia Production, Global Industry Benchmark Report,* http://www.fertilizer.org/imis20/images/Library_Downloads/2014_ifa_ff_ammonia_emissions_july.pdf?WebsiteKey=411e9724-4bda-422f-abfc-8152ed74f306&=404%3bhttp%3a%2f%2fwww.fertilizer.org%3a80%2fen%2fimages%2fLibrary_Downloads%2f2014_ifa_ff_ammonia_emissions_july.

Kannan R. and W. Boie (2003), Energy management practices in SME - Case study of a bakery in Germany, *Energy Conversion and Management*, Vol. 44, No. 6.

Kendall, A., E. Marvinney, S. Brodt and W. Zhu (2015), "Life Cycle–based Assessment of Energy Use and Greenhouse Gas Emissions in Almond Production, Part I: Analytical Framework and Baseline Results", *Journal of Industrial Ecology*, Vol. 19, No. 6.

Känkänen, H. (2015). "Reducing use of fossil energy by biological N fixation", *Legume Perspectives*, No. 8.

Lampkin, N.H. (2007), "Organic farming's contribution to climate change and agricultural sustainability", paper presented at the Welsh Organic Producers' Conference, 18 October.

Lincoln University (2006), *Comparative Energy/Emissions Performance of New Zealand's Agriculture.*

Martin-Gorriz, B., M. Soto-García and V. Martínez-Alvarez (2014) "Energy and greenhouse-gas emissions in irrigated agriculture of SE (southeast) Spain - Effects of alternative water supply scenarios", *Energy*, Vol. 77.

Meyer-Aurich A., A. Schattauer, H. Hellebrand, H. Klauss, M. Plöchl and W. Berg (2012), "Impact of uncertainties on greenhouse gas mitigation potential of biogas production from agricultural resources", *Renew Energy*, Vol. 37.

Monforti-Ferrario, F., J.-F. Dallemand, I. Pinedo Pascua, V. Motola, M. Banja, N. Scarlat, H. Medarac, L. Castellazzi, N. Labanca, P. Bertoldi, D. Pennington, M. Goralczyk, E. M. Schau, E. Saouter, S. Sala, B. Notarnicola, G. Tassielli and P. Renzulli (2015), *Energy use in the EU food sector: State of play and opportunities for improvement*, European Commission, Joint Research Centre, Brussels, Belgium.

Nguyen, T., J. Hermansen and I. Mogensen (2010a), "Fossil energy and GHG saving potentials of pig farming in the EU", *Energy Policy*, Vol. 38.

Nguyen, T., J. Hermansen, I. Mogensen (2010b), "Environmental consequences of different beef production systems in the EU", *Journal of Clean Production*, Vol. 18.

National Farmers' Union (NFU) (2010), "Kemble Farms Ltd: Focus on Anaerobic Digestion", www.farmingfutures.org.uk/resources/case-studies/kemble-farms-ltd-focus-anaerobic-digestion.

Natural Resources Defence Council (NRDC) (2009), *Water Efficiency Saves Energy: Reducing Global Warming Pollution Through Water Use Strategies.*

OECD (2016a), *Farm Management Practices to Foster Green Growth*, OECD Green Growth Studies, OECD Publishing, Paris, DOI:10.1787/9789264238657-en.

OECD (2016b), "Synergies and Trade-offs between Agricultural Productivity and Climate Change Mitigation and Adaptation: Dutch Case Study", COM/TAD/CA/ENV/EPOC(2016)7/FINAL.

OECD (2016c), *World Energy Balances 2016*, OECD Publishing, Paris, http://dx.doi.org/10.1787/9789264263116-en.

Office fédérale de l'agriculture (OFAG) (2015), *Rapport agricole,* Bern, Switzerland, http://2015.agrarbericht.ch/fr.

Pelletier, N., E. Audsley, S. Brodt, T. Garnet, P. Henriksson, A., Kendall, K. Kramer, D. Murphy, T. Nemecek, M. Troell and P. Tyedmers (2011), "Energy Intensity of Agriculture and Food Systems", *Annual Review of Environmental Resources*, Vol. 36, No. 7.

Pelletier, N., R. Rasmussen and R. Pirog (2010), "Comparative life-cycle impacts of three beef production strategies in upper Midwestern United States", *Agricultural Systems*, Vol. 103, No. 6.

Pimentel, D. (2009), "Reducing energy inputs in the agricultural production system", http://monthlyreview.org/2009/07/01/reducing-energy-inputs-in-the-agricultural-production-system.

Pimentel, D. (2006), *Impacts of organic farming on the efficiency of energy use in agriculture*, The organic Centre, Cornell University.

Profita, C. (2012), "Which is greener: Grass-fed or grain-fed beef?", October 2012, www.opb.org/news/blog/ecotrope/which-is-greener-grass-fed-or-grain-fed-beef/

Rusu, T. (2014), "Energy efficiency and soil conservation in conventional, minimum tillage and no-tillage", www.sciencedirect.com/science/article/pii/S2095633915300575.

Sands, R., P. Westcott, J. Price, J. Beckman, E. Leibtag, G. Lucier, W. McBride, D. McGranahan, M. Morehart, E. Roeger, G. Schaible and T. Wojan (2011), *Impacts of Higher Energy Prices on Agriculture and Rural Economies*, Economic Research Service, United States Department of Agriculture, ERS Report No. 123, August 2011.

Schader, C. (2009), "Cost-effectiveness of organic farming for achieving environmental policy targets in Switzerland", FiBL, Frick, and Aberystwyth University.

Schneider, U. and P. Smith (2009), "Energy intensities and greenhouse gas emissions in global agriculture, *Energy Efficiency*, Vol. 2, Issue 2.

Smith, L., A. Williams and B. Pearce (2015), "The energy efficiency of organic agriculture: A review", *Renewable Agriculture and Food Systems*, Vol. 30, Issue 3.

Smith, E., R. Zentner, C. Nagy, M. Khakbazan and G. Lafond (2008), "Decoding your fuel bill: What is your farm's real energy bill?", http://prairiesoilsandcrops.ca/articles/volume-1-5-print.pdf.

Tuomisto, H., I. Hodge, P. Riordan, D. Macdonald (2012), "Does organic farming reduce environmental impacts? – A meta-analysis of European research", *Journal of Environmental Management*, Vol. 112, pp. 309-320.

Van Ierland, E. and A. Lansink (2003), *Economics of Sustainable Energy in Agriculture*, Kluwer Academic Publishers.

Vert J. and F. Portet (co-ord.) (2010), *Prospective Agriculture Énergie 2030 - L'agriculture face aux défis énergétiques*, Centre d'études et de prospective, SSP, Ministère de l'Agriculture, de l'Alimentation, de la Pêche, de la Ruralité et de l'Aménagement du Territoire, France.

Woods J., A. Williams, J. Hughes, M. Black and R. Murphy (2010), "Energy and the Food System", *Royal Society*, Vol. 365, No. 1554.

Xiarchos, I.M. and B. Vick (2011), "Solar Energy Use in US Agriculture: Overview and Policy Issues", United States Department of Agriculture, Office of the Chief Economist, Office of Energy Policy and New Uses.

Annex 2A.1

Empirical evidence on direct on-farm energy use and efficiency

Table A2.1. Direct on-farm energy efficiency, 1995-97 and 2010-12 (1990=100)

	1995-97	2010-12	Change (%)
OECD			
Australia	99	76	-24
Austria	108	120	12
Belgium	105	92	-13
Canada	86	71	-17
Czech Republic	89	143	62
Denmark	98	103	5
Estonia	319	294	-8
Finland	114	109	-4
France	96	80	-17
Greece	119	223	88
Hungary	125	158	26
Iceland	90	121	33
Ireland	86	103	19
Israel	69	64	-8
Italy	107	111	4
Japan	77	105	36
Korea	56	85	50
Latvia	163	156	-4
Luxembourg	85	54	-36
Mexico	107	98	-9
Netherlands	89	117	32
New Zealand	99	103	4
Norway	66	139	112
Poland	56	71	29
Portugal	89	133	50
Slovak Republic	154	203	32
Slovenia	95	84	-12
Spain	78	81	4
Sweden	92	135	47
Switzerland	50	98	96
Turkey	74	55	-26
United Kingdom	97	132	37
United States	102	97	-5

Table A2.1. Direct on-farm energy efficiency, 1995-97 and 2010-12 (1990=100 (*cont.*)

	1995-97	2010-12	Change (%)
Non-OECD			
Argentina	72	101	40
Brazil	107	145	36
China	136	195	43
Colombia	99	101	2
Costa Rica	30	63	112
India	69	75	8
Indonesia	78	75	-3
Philippines	157	115	-27
Russian Federation	90	198	121
South Africa	66	109	65
Ukraine	142	212	49
Viet Nam	85	122	43
Groupings			
OECD Total	94	99	6
Non-OECD	143	200	40
EU-28	110	140	28
G7	95	93	-3
G20	102	127	24
World	61	76	26

Table A2.2. Composition of direct energy use in the OECD area, 1990-2013 (%)

	1990-94	1995-1999	2000-04	2005-09	2010-13
		Agriculture			
Coal and coal products	3.5	2.7	2.0	1.9	2.0
Oil products	75.7	76.9	72.2	68.6	68.2
Natural gas	8.2	7.6	9.2	10.2	8.9
Biofuels and waste	1.7	2.0	2.1	2.7	3.8
Electricity	9.6	9.8	13.5	15.8	16.0
		Food processing and manufacture			
Coal and coal products	14.2	8.3	8.1	8.6	8.5
Oil products	26.8	21.2	18.0	14.4	8.9
Natural gas	21.8	39.0	40.7	41.2	45.8
Biofuels and waste	5.1	5.6	5.8	5.1	5.5
Electricity	29.2	24.5	25.9	28.6	28.9
Heat	3.0	1.5	1.5	2.1	2.3

1. Data are based on ISIC REV.4, Division, 10, 11 and 12. They include processing and preserving of fish, crustaceans and molluscs.
2. Primary and secondary oil includes crude and refined oil products.
Source: IEA (2016), *World Energy Statistics and Balances,* online data service 2016 edition, http://dotstat.oecd.org/?lang=en

Table A2.3. Summary of direct energy uses by activity and sector in English agriculture

Category	Energy use	Fuel type
Indoor horticulture	Heating (c. 90% for most crops)	Natural gas, oil, coal, and liquid propane gas (LPG)
	Lighting	Electricity – can be a significant demand to extend "natural" growing periods
	Pumping (irrigation and water circulation for temperature regulation and hydroponic cultivation)	Electricity
	Ventilation	Electricity
Field horticulture	Field operations: tractors, crop treatment machinery, harvesting	Diesel (gas oil)
	Crop cooling and storage	Electricity – increasing demand for year-round storage
	Crop preparation	Electricity
Intensive livestock (pigs, poultry meat and eggs)	Heating	Oil and LPG, some electricity
	Ventilation	Electric fans
	Lighting	Electricity
	Pumping and materials handling	Electricity
Dairy	Milking machinery	Electric pumps
	Milk cooling	Electricity
	Water heating	LPG fuel boilers
	General lights and power	Electricity
	Field machinery	Diesel
	Manure management	Diesel and electricity
Beef and sheep	Lights and power	Electricity
	Forage production	Diesel
Arable crops	Field operations	Diesel (gas oil)
	Grain drying, storage and handling	Electricity/gas/oil/LPG
Potatoes and sugar beet	Field operations	Diesel in tractors
	Crop cooling and storage	Electricity
	Crop preparation	Electricity

Source: DEFRA (2010).

Table A2.4. The most cost-effective measures and net savings or costs

Category	Net savings (negative value in brackets) and cost (positive value in brackets) GBP per t of CO_2 abated were:
Arable	Improved control of grain drying facilities (-115); Minimum tillage (-108); Improved management (-68); Improved maintenance (-13); Variable speed drives in crop drying facilities (77).
Indoor horticulture: High-temperature edibles	Improved energy management (-174); High efficiency boilers (-166); System insulation (-161); Biomass boilers (-134); Climate control computers (-126); Improved maintenance (-120); Heat recovery (-73); Variable speed drives (-53); Improved cooling systems (14); Improved light sources (62).
Indoor horticulture: Low temperature edibles	Improved energy management (-177); Air leakage minimisation (-162); Biomass heaters (-158); Improved maintenance (-154); High efficiency boilers (-151); System insulation (-118); Variable speed drives (-26); Climate control computers (9). Improved cooling systems (14); Improved light sources (62).
Poultry meat	Automatically controlled natural ventilation (-324); High efficiency. lighting (-209); Improved energy management (-190); Installation of biomass boilers (-144); Destratification (-96); Insulation/sealing of buildings (-83); Improved maintenance (-56); Improved insulation of ventilation systems (-38); Heat recovery (7).
Eggs	ACNV (Automatically controlled natural ventilation) (-271); Improved energy management (-201); Installation of high-efficiency lighting (-152); Improved maintenance (-32); High-performance fans (27).

(continued on next page)

Table A2.4. The most cost-effective measures and net savings or costs *(continued)*

Category	Net savings (negative value in brackets) and cost (positive value in brackets) GBP per t of CO_2 abated were:
Dairy	Control optimisation (-259); Adoption of heat recovery (-236); Variable speed drives (-215); Optimisation of pre-cooling (-210); Biomass boilers (-177); Improved energy management (-169); Improved maintenance (-125); Optimisation of insulation (-95); Converting muck-scrapers to electrically-powered units (-54); High-efficiency lighting (-11).
Beef and Sheep	No cost-effective measures identified for sheep and limited potential in beef
Pigs	Improved control of weaner heating (-220); Installation of biomass boilers for creep heating (-209); Improved insulation of weaner buildings (-198); Installation of biomass boilers for weaner heating (-193); Insulation of creep buildings (-185); Under-floor creep heating (-132); Improved energy management (-108); ACNV (-103); Improved control of creep heating (-63); Heat recovery in weaner buildings (-6); Improved maintenance (15).

Source: DEFRA (2010).

Chapter 3

Energy use and energy efficiency opportunities in the downstream food chain

Drawing on existing information which is readily available for OECD and some non-OECD countries, this chapter reviews the current state of knowledge with respect to energy use and efficiency beyond the farm – food processing, food retailing restaurants and catering. Key drivers and trends in the energy use and efficiency are highlighted, along with opportunities for and constraints upon improved efficiencies.

The food processing industry

Food processing requires a substantial amount of energy

This part of the food chain makes a significant, and increasing, contribution to the GHGs emitted by the food chain and its overall energy consumption. Evidence from the United States suggests that this growth is a result of households outsourcing food preparation to processors (Canning et al., 2010). Moreover, the study indicates that the processing expansion to meet this demand is partly due to changing relative prices arising from: 1) the changing opportunity cost of time spent in home food preparation relative to the cost of prepared food; 2) until recently, stable energy prices, relative to increasing labour costs in the processing industry.

The food processing industry is, for example, the fifth biggest consumer of energy in the United States (Wang, 2009). It is also a sizeable consumer of energy within the agro-food chain, representing 28% of energy embedded in consumed food in the European Union, for example (Monforti-Ferrario et al. 2015). Of the total energy used by the food industry, process heat for thermal processing and dehydration consumes approximately 59%, while refrigeration uses approximately 16% and motor drives 12% (Wang, 2009).

Relative to the complexity of farm level energy uses, this stage comprises more direct energy use. In OECD countries, on average, total direct energy consumed by the food industry, amounted to 64.5 Mtoe fuel equivalent, accounting for about 2% of the OECD average final energy consumption in 2011-13. This this share broadly ranged between a few tenths of percentage points and 6% of the national final energy consumption, with New Zealand, Denmark, Ireland and Australia having the larger shares (Figure 3.1). Natural gas has dominated this sector's energy mix in 2011-13, followed by electricity, coal and oil (Figure 3.2).

As for any other industrial product, the more processed the food, the higher the energy consumption required. The energy consumed per unit of processed foodstuff is very diverse among products and even for the same product can be very different, depending on the country.

Market competitiveness remains the strongest driver for energy efficiency solutions

The food processing and beverage sector has been improving its energy intensity over a long period of time. These moves have primarily driven by overall market conditions, growth and competitiveness. Food and beverage processing energy efficiency can help food and beverage producers stay competitive by increasing production while shrinking their energy costs. The sector has demonstrated a good track record in improving energy efficiency (Figure 3.3).

Numerous studies have focused on energy needs for processing particular food products.[1] Overall, a positive trend is emerging as a number of sectors are beginning to report improvements in energy efficiency in recent years. The food and drink industry in Europe has reported an increase in energy efficiency and decrease of energy GHG emissions per unit of production value, suggesting an ongoing trend towards a more appropriate and optimised energy use in the sector (FoodDrinkEurope, 2012). In the European Union, for example, energy consumption in the food industry has steadily decreased in recent years, both in absolute terms and, even more, in terms of energy consumption per unit of production value (Monforti-Ferrario et al. 2015). The detailed study by Ramírez, Patel and Blok (2006) on the Dutch food processing industry based on a much more complex pool of indicators, confirmed the trend. Fonterra, a farmer-owned co-operative and the largest milk processor in New Zealand, has reduced its energy consumption by 13.9% per tonne of product since 2003 (Federated Farmers of New Zealand, 2015).

In the EU28, the food and beverage sector accounts for 10% of the energy consumed by the industrial sector, which accounts for a quarter of total EU28 final energy consumption. Process heating is the most significant source of energy use. Natural gas consists of almost half of the sector's final energy requirement (45%), followed by electricity (34%). Energy consumption fluctuates year-on-year primarily due to the production fluctuation in the wide range of product scope within the sector, in which the specific energy intensity and energy consumption is fully dependent on the product output type and the associated processes in manufacturing the finished product to the specification as dictated by the consuming market on the specific year. Gradual improvement in energy efficiency is observed over the period 1995-2012, during which energy consumption was reduced by approximately 0.6% despite an increase of 85% in GDP – primarily driven by innovation of new products and quality (EC, 2015).

Figure 3.1. A substantial amount of energy is consumed in food processing and manufacture

Share in total energy consumption, 2011-13 (%)

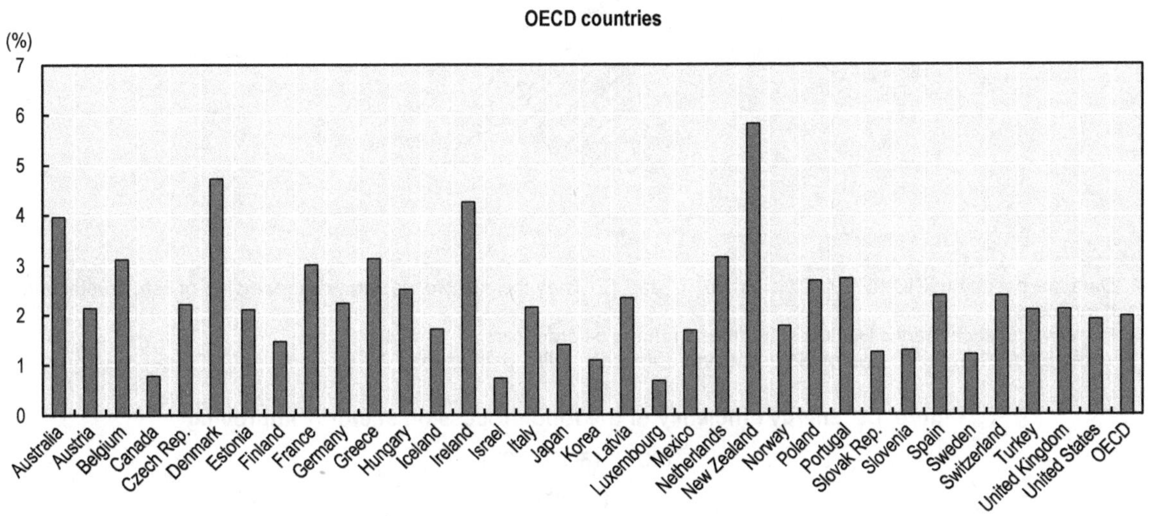

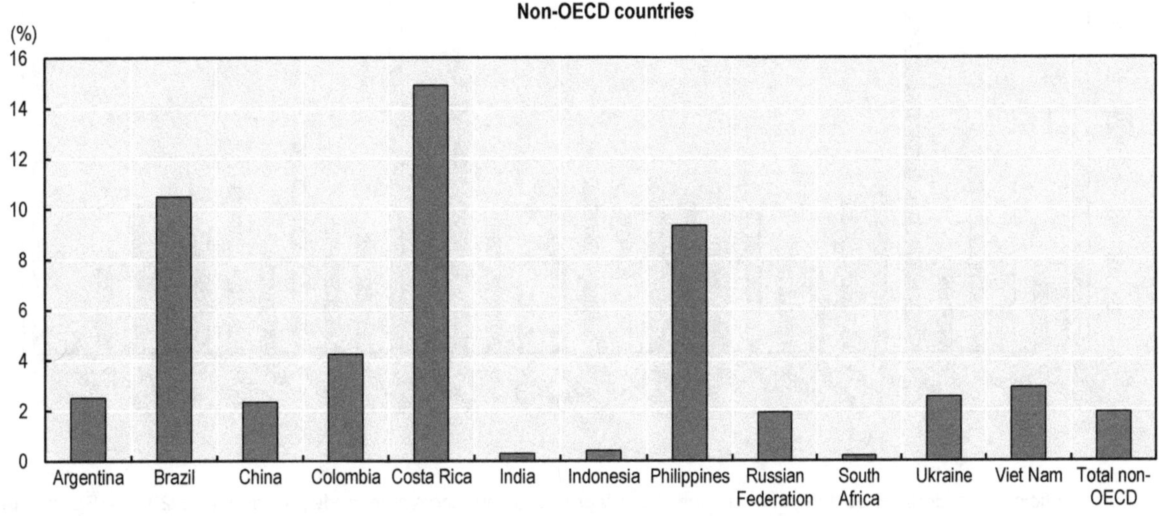

Note:
1. Data are based on ISIC REV.4, Division, 10, 11 and 12. They include processing and preserving of fish, crustaceans and molluscs.
Source: IEA (2016), *World Energy Statistics and Balances,* online data service 2016 edition, http://dotstat.oecd.org/?lang=en.

Figure 3.2. Natural gas is the main source of energy in food processing and manufacture in the OECD area,

(2011-13, %)

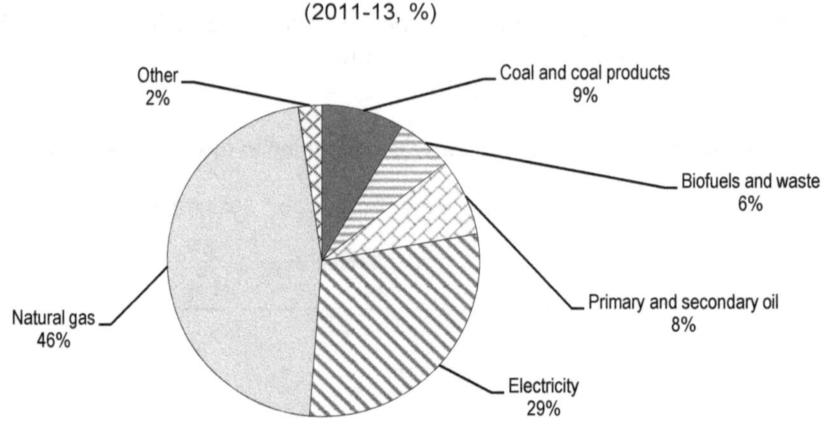

Notes:
1. Data are based on ISIC REV.4, Division, 10, 11 and 12. They include processing and preserving of fish, crustaceans and molluscs.
2. Primary and secondary oil includes crude and refined oil products.

Source: IEA (2016), *World Energy Statistics and Balances,* online data service 2016 edition, http://dotstat.oecd.org/?lang=en.

Figure 3.3. Energy efficiency of the food processing sector is improving

(2011-13/1995-97)

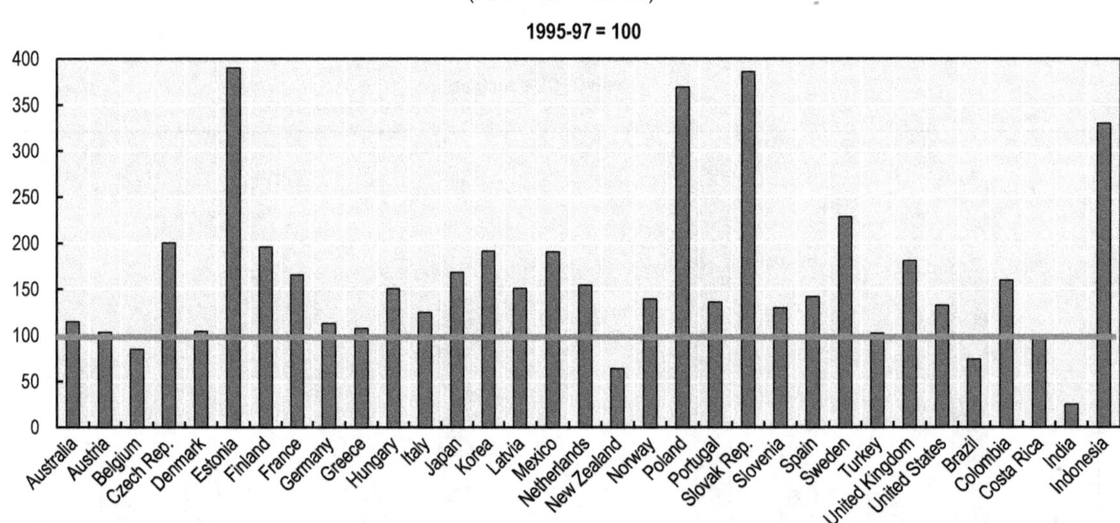

Note:
Energy efficiency is defined as the ratio of food, beverages and tobacco value added in constant 2010 USD per unit of energy use.

Source: IEA (2016), *World Energy Statistics and Balances,* online data service 2016 edition, http://dotstat.oecd.org/?lang=en.

In the United Kingdom, beverages, dairy products and meat processing are the largest parts of the food and drink processing industry (DEFRA, 2007; 2013). Fossil fuel for use in boilers that supply steam for the process is the main energy source used (49%). Also high is the use of energy for direct heating using fuel (19%). Electricity is predominantly used for the processes involving compressed air, refrigeration and for other processes that use motors for mixing and stirring. Technological improvements have been implemented in the food manufacturing sector, resulting in a general trend of reduction in carbon dioxide emissions since

1990, but efficiency enhancements have to continue in order to lead to a competitive and sustainable food sector (Campden BRI, 2011).

Broad benefits of energy use reduction and energy recovery from processing wastes

Many food processing operations, such as sterilisation, pasteurisation, cooking, dehydration, drying and freezing, involve the transfer of heat into or out of foods to improve eating quality and safety or to extend shelf life – an important factor in food waste reduction. Dehydration and drying are particularly energy intensive, due to the relatively low energy efficiency of industrial dryers (Wang, 2009). A number of methods have been applied to increase the energy efficiency of the dehydrating and drying process such as mechanical processes (filtration and centrifugation) and multiple-effect evaporator system, which collects and reuses vapour in order to reduce energy requirements for the removal of product moisture. However, potentially energy efficiency gains will need to be weighed against the cost of installing additional evaporators.

Energy-efficient drying technology in the sugar industry

By using mechanical screw presses to remove as much water as possible from sugar beet pulp before drying, the British Sugar Beet factory at Wissington, in the United Kingdom, decreased the energy consumed by its dryers, saving 55.8% in primary energy use (Best Practice Programme, 1997, in Wang, 2009). As a result of measures such as these, between 1990 and 2009 the company achieved a 25% reduction in the amount of energy used to produce a tonne of sugar. The company is currently seeking to achieve a 30% reduction in the amount of energy it uses to produce a tonne of sugar by 2020, as measured against the same 1990 baseline (British Sugar, 2010).

Where possible, the most cost-effective use of heat is usually to improve the energy efficiency of the heat generating process itself. The economic benefits of waste heat recovery include the reduction of energy costs and of capital costs for energy conversion equipment with less capacity. However, the economics of a waste heat recovery system depends on the use, quantity and quality of the recovered waste heat, and the heat transfer equipment for waste heat recovery. In the majority of cases, heat recovery is far more efficient when the heat source and the use of the heat are located close together and are available at the same time (Carbon Trust, 2011).

Increasing waste heat recovery in chocolate manufacturing

In 2010, food and drinks company Nestlé upgraded the coal-fired plant at its chocolate factory at Halifax, in the United Kingdom, to a system that was able to trap waste heat from refrigerating chocolate and then convert it in order to shape chocolate products. The new system, which reportedly has a 15% higher energy efficiency rate than the previous system, has enabled the factory to reduce its CO_2 footprint by 1.1 million pounds annually, and save the company almost USD 400 000 in energy costs a year (Kaye, 2013).

The cogeneration of steam and electricity has become a common means in the international sugarcane industry to significantly increase boiler efficiency (Wang, 2009). High-pressure steam is first used to generate electricity, after which the exhaust steam is used again as processing steam. Because of the dual use of steam, the overall energy efficiency of the combined steam and electricity system in sugar refineries is 75-80% compared to the 35% of electricity generation alone. Cane sugar refineries consequently generate the majority of their own electricity.

Many European food and drink manufacturers have set out internal goals to reduce their environmental footprints and a large number are certified or in the process of being certified with ISO 14001 (International Standard on environmental management systems) or the EU Eco-Management and Audit Scheme (EMAS). The European Food and Drink industry currently has the fifth highest number of EMA registered organisations in any sector (FoodDrinkEurope, 2012). Annex 3A.1 highlights many of the changes that have already been put place in recent years and some proposed changes that will reduce energy consumption and improve energy efficiency by the FoodDrinkEurope.

In the United Kingdom, energy efficiency of the food and drink manufacturing industry is mainly addressed through the Climate Change Agreements (CCA) policy framework (see Chapter 6). Over the period 1999-2010 the Food and Drink Federation CCA (which covers about half of the food and drink manufacturing sector) reported a 21% improvement in energy efficiency and 18% decline in energy use levels. From an environmental perspective CO_2 emissions associated with this energy use fell by 22% over the same period.

In the United Kingdom, energy consumption figures also indicate that a small number of products are responsible for 80% of the GHG emissions (FDF, 2010). The most prominent of these are manufacture of bread and fresh pastry goods, production of cheese and other dairy products, production of meat and poultry products, and manufacture of beer and alcoholic beverages. This therefore points to the need for improvements in technologies used by these sectors, such as processing equipment, refrigeration, boilers, ovens, pumps, space heating and lighting.

Combined heat and power plants reducing fuel consumption in the sugar industry

At British Sugar factories, coal, oil or gas are used to fuel water boilers which produce the steam needed for electricity generation for the factory. The steam is used again in the evaporation stages, and later to heat the sugar juice throughout the process. At several factories, more electricity is generated than the factory requires. In 2008, in addition to meeting over 94% of its own electricity requirements, British Sugar's CHP plants generated an additional 700 000 MW hours of electricity which for export to the local electricity network. Combined cycle gas turbines have been installed at two factories in the United Kingdom, enabling the company to extract around 80% of the energy contained within the fossil fuel used during a production campaign – double the amount extracted by a conventional power station. This substantially reduces fuel consumption and associated CO_2 emissions (British Sugar, 2010).

The food retailing sector

Refrigeration and food transport are the main energy users

Distribution and retailing involve a range of energy uses, including in particular refrigeration and food transport. The energy consumption in food retailing varies widely and depends on many factors such as the type and size of the store, business practices and refrigeration and environmental control systems used.

Although the energy consumption (and CO_2 emissions) vary widely between establishments and practices, the key energy using activities in distribution and retail are energy management and monitoring; transport; chilled storage (in transport vehicles, warehouses and retail facilities); ventilation and air-conditioning; lighting; conveying, forklifts, etc.; chilled storage; lighting, and cooking (supermarket bakeries, etc.).

Overall, the need for more refrigeration in the food chain reflects the growing length of the food chain and the distance between producers and consumers in terms of both space and time (i.e. the availability of out of season produce), and increased health concerns and regulations. The refrigeration systems in the United Kingdom account for between 30% and 60% of the electricity used at this stage, whereas lighting accounts for between 15% and 25% with the heating, ventilation and air-conditioning equipment and other utilities such as bakery, for the remainder (Tassou et al., 2014).

The Tassou, Hadawey and Marriott (2011) study on energy consumption of 2 570 retail food stores in the United Kingdom found a wide variability of energy intensity even within stores of the same retail chain. If the electrical intensity of the stores above the average is reduced to the average by energy conservation measures, annual energy savings of the order of 10% (representing 355 000 tonnes reduction in CO_2 emissions per annum) can be achieved.

The food industry is transport intensive.[2] Energy use and efficiency vary significantly according to the mode of transport used, as well as the volume of products transported per journey and the use of refrigeration. Long distance transport by air transport has relatively high energy inputs, but only around 1% of food products are shipped by air, with around two-thirds of local products and one-third of exported products by road, and the remainder fairly equally divided between rail, shipping, and local waterways (FAO, 2011).

Typically, the energy input for transport is a relatively small share of total energy inputs into an agri-food chain. Therefore producing specific crops and animal products in locations where productivity is naturally higher due to soil and climatic conditions can sometimes outweigh any possible transport savings if grown locally but yielding less.

Various approaches can be employed to reduce transport emissions and increase energy efficiency; these include: food chain optimisation and reduction in food transport intensity; modal shift; improved vehicle fuel efficiency; alternative refrigeration technologies and electrification of transport with zero or low-carbon electricity (Tassou et al., 2014).

Discussion of transportation tends to raise conflicting evidence on the benefits of local food. A central notion behind this rising demand is that local food production is less energy – and emissions – intensive than larger scale and more centralised production. This is due to perceived shorter transport routes that food is transported from the point of production to the consumer (lower "food miles") and also perhaps due to local production using smaller scale, less intensive production plants.

While at first glance the local sourcing of produce closer to the point of production (e.g. farmers' markets or local grocery establishments) would appear to reduce distribution fuel use and transport costs, a range of scale factors may in fact lead to a counteracting increase in resource use and efficiency. In the process of economic development economies of scale are created precisely by minimising resource use per unit of output, and hence corresponding to the objective of resource use efficiency. The consumption of locally-produced foods is therefore not necessarily more energy efficient, particularly when these foods are consumed out of season (Edwards-Jones et al., 2008; Sim et al., 2007; DEFRA, 2008).

Pretty et al. (2005) compare the estimated costs of food production externalities in United Kingdom agriculture to the costs of externalities associated with the transport of food to retail outlets and transporting food to consumers' homes. When these other aspects are included, their estimate of the total cost of externalities associated with food consumption increases from GBP 1.5 billion (production externalities) to GBP 5.2 billion (production and delivery externalities). The authors note that even this higher estimate may be too low given that they do not take account the externalities generated from energy consumption by processors, manufacturers and wholesalers for light, heat, refrigeration and transport, the disposal of food packaging, foods consumed by domestic pets, methane emissions from landfills and sewage waste, and the energy required for domestic food preparation. The estimated cost associated with transport from farm to retail outlets alone is GBP 2.3 billion or 55% higher than the estimated costs of production externalities.

Pretty et al. (2005) use a broad definition of external costs imposed by vehicle transport in their analysis, which includes costs imposed by congestion, harm to health (noise, asthma), climate change (from greenhouse gases) and infrastructure damage. In contrast, Weber and Matthews (2008) focus on the life-cycle GHG emissions associated with food production compared to long-distance distribution in the United States. The authors find that food is transported long distances in general, with the average delivery distance is 1640 km with 6760 km for the life-cycle supply chain. However GHG emissions associated with food are dominated by the production phase rather than distribution. They estimate that transportation as a whole represents 11% of the CO_2 footprint of the average US household's consumption of food, compared to 83% for production. The authors also find that the various food groups exhibit significant differences in GHG intensity. For example, red meat is 150% more GHG-intensive than chicken or fish. On this basis the authors conclude that dietary shifts (substituting chicken and fish for red meat) can be more effective than local sourcing of food if the aim is to reduce the food-related climate footprint.

The analysis by Saunders and Barber (2008) of life cycle assessment comparative energy efficiency in United Kingdom imports of agricultural products from New Zealand to those produced domestically supports the conclusion reached by Weber and Matthews that transportation is a relatively small contributor to total energy use in agriculture and that local sourcing of food may not necessarily lead to a reduction in total energy use and GHG emissions. Because they determine that the production of lamb is more intensive in terms of energy and other inputs in the United Kingdom than in New Zealand, they estimate that energy used (and GHG emissions) per tonne of product are more than four times larger for lamb produced in the United Kingdom than for lamb imported from New Zealand (calculated on a carcass weight basis), despite the fact that transportation from New Zealand to the United Kingdom accounts for 18% of the emissions for the New Zealand product. They also estimate that imported dairy products yield roughly half the energy (and emissions) of domestically produced dairy products (calculated on a milk solids basis). In the case of apples, the New Zealand source was 10% more energy efficient. In case of onions, whilst New Zealand used slightly more energy in production, the energy cost of shipping was less than the cost of storage for out-of-season consumption in the United Kingdom.

Williams et al. (2007) examined seven products for the United Kingdom market. Their report concluded that food imports from countries where productivity is greater or where refrigerated storage requirements are lower could have a smaller carbon footprint than locally produced food.

While these studies can be criticised on various grounds – their methodology, assumptions used and lack of hard data in some cases, it seems clear that the concept of food miles as a guide to policy is questionable. Simply because a product is transported a long distance from the point of production to point of consumption does not imply that it will generate a larger environmental footprint than a product that is produced and consumed locally. The environmental impact of food production and consumption is often difficult to determine and depends as much on the behaviour of consumers as it does on the actions of producers, processors and others who form part of the food system.

Retailers are committed to increasing energy efficiency

Like food manufacturing, the retail sector has acted to reduce energy-related carbon emissions and improve energy efficiency. For example the British Retail Consortium (which includes non-food retailers) reports that since 2005, signatories to *A Better Retailing Climate* launched in 2008 have: achieved annual improvements of 2-3% in their energy efficiency (as measured in energy consumption per unit of floor area); cut energy-related emissions from buildings by 30%; cut GHGs from super market refrigeration by 55%; cut energy-related carbon emissions from store deliveries by 29%; and cut the proportion of waste sent to landfill from 47% to 6% (BRC, 2014).

By 2020 signatories have committed (compared with the 2005 base) to reduce absolute carbon emissions from retail operations by 25%; cut energy-related emissions from buildings by 50%; reduce emissions from refrigeration gases by 80%; reduce carbon emissions from store deliveries by 45%; and send less than 1% of their waste to landfill. In addition the retailers are working with suppliers to increase the sustainability of the products they sell; and this could have a more substantial impact than the savings from retailing alone given the power of retailers within modern food chains and notably their ability to influence the production or manufacture of private label products (products for sale under supermarkets' own brands).

... but there is no silver bullet to improve energy efficiency

According to the BRC (2014) innovations that have been made to improve energy efficiency since the mid-2000s by retailers include:

- *Retail operations:* improving energy monitoring and control systems; and improving the operational efficiency through placing doors on fridges and chillers and implementing auto-defrost processes to tackle waste energy consumption.

- *Energy use in building:* up-scaling deployment of energy-efficient technologies such as light-emitting diode (LED) lighting, trialling new and innovative technologies in refrigeration, heating and ventilation equipment and increasing the use of renewable energy on site such as biomass boilers, solar power and wind turbines.
- *Transport:* increasing the proliferation of alternative fuels in fleets, such as bio-diesel and fuels from waste and using more efficient vehicles (including dual fuel tractors); developing better route optimisation models; increasing delivery efficiency; and driving techniques.
- *Human capital:* Tracking energy consumption through behaviour change and staff training programmes.

These findings suggest that the improvements that have been achieved are best understood as the outcomes from a range of disparate activities and actions (covering capital expenditures, improved monitoring and control systems, training, and so forth) directed at reducing energy use and GHG emissions. In particular, the following common themes emerge as to how have these companies managed to deliver these efficiency gains:

- *Having the infrastructure for energy management is essential:* This encompasses aspects such as clearly defined responsibilities, robust data, on-going monitoring and performance review. Of particular importance in this regard is the setting and publishing of targets. These create expectations and accountabilities and mean that the organisational focus on energy efficiency is maintained over time.
- *Setting targets relating to the development of innovations and to the wider deployment of innovative technologies and approaches to energy efficiency across the entire business.* For example, in relation to transport, retailers have to work with engine and vehicle manufacturers to test a range of new technologies to ensure that cost-effective innovations are deployed across the entire business. Having targets relating to the development of innovations and to the wider deployment of innovative technologies and approaches across their business might be necessary.
- *Investments in energy efficiency should yield a reasonable return on investment,* with the vast majority of the actions that have been taken to reduce energy consumption being justified in cost-benefit terms. Moreover, the retailers expect their focus on energy management to be sustained because they expect energy prices to remain high over the long term.

Energy efficiency on its own is just one part of the wider discussion around corporate climate change performance. Obviously there could be conflicts between business growth, and energy and emissions reductions, and retailers often do not report on their activities or performance due to confidentiality. Yet evidence suggests that there is much that companies can achieve through explicitly focusing on improving energy efficiency and reducing GHG emissions. However, achieving these outcomes requires a systematic and consistent focus on energy efficiency, underpinned by robust management and monitoring systems and processes.

Restaurants and catering

Catering operations consume substantial amounts of energy

This sector is extremely diverse, encompassing large contract caterers, the fast food industry, hotels, restaurants and non-commercial catering such as schools and hospitals. For the larger groups, opportunities for savings are similar to those in manufacturing and retailing, including lighting, ventilation and more efficient cooking and refrigeration equipment (Tassou et al., 2013; Cibse, 2009) and transport. However, behavioural changes with respect to type of food consumed, food preparation practices, environmental conditions in the catering premises and their interaction with refrigeration and Heating, Ventilation and Air Conditioning (HVAC) equipment provide additional energy-saving opportunities.

Energy use by catering facilities normally accounts for 4-6% of these facilities' operating costs. Given that the profit margin of many caterers is within this range, energy saving can directly increase revenue and profitability without the need to increase sales. The Carbon Trust (2012) argues that with moderate improvements in efficiency and the effective use of equipment, energy savings of up to 20% are achievable in the United Kingdom catering industry, for example, generating overall savings for the sector of over GBP 80 million per year.

Food packaging can represent a significant potential for energy efficiency gains in the agro-food chain

The growth in the production of processed, convenience and take-away foods in recent years has led to a rise in the use of food packaging (Environment Agency, 2010). Food packaging can represent a significant potential for energy-efficiency gains in the agro-food chain and the food companies, retailers and the packaging sector are investing significantly in innovation in environmentally-friendly packaging materials and in design to enable packaging reduction, recycling and reuse. The energy implications of different packaging materials depends on a number of factors across their lifecycle, including their weight, recyclability, degradability, energy feedstocks required for their manufacture, and so on.

Cutting energy use by changing product packaging

US retailer Wal-Mart partnered with beverage company The Wine Group to reduce the weight of its Oak Leaf brand wine bottles in order to reduce waste and cost. Weighing 37% less than the previous bottle, the new bottles reduced the overall weight of packaging by more than 27%. Additionally, the changes resulted in the need for 280 fewer trucks to transport the bottles and therefore reduced the carbon footprint of the product. Walmart was also able to reduce the retail price by nearly 7% (Packaging Digest, 2012).

But, lighter packaging can influence negatively food waste and generate energy waste in other parts of the chain. Wrapped cucumbers, for example, have been found to maintain their freshness for 11 days longer than those which are unpackaged, which is likely to translate into reduced food waste: 16% of packaged fruit and vegetables are reportedly wasted at the retail and household levels, compared with 32% of non-packaged produce, for example (FoodDrinkEurope, 2012).

Notes

1. Examples of recent studies include: almonds (Kendall et al., 2015); milk (Mancini, 2011; Ramirez, Patel and Blok, 2006a), Parmesan cheese (Mancini, 2011), meat (Fritzson and Berntsson, 2006; Mancini, 2011; Ramirez, Patel and Blok, 2006b), pasta (Mancini, 2011), pastry (Kannan and Boie, 2003), rice (Mancini, 2011) and natural orange juice (Mancini, 2011).

2. In France, for example, food transport represents around 28.8% of total industry transport in kilometres per tonne, whilst in the United Kingdom it is estimated to account for 25% of all HGV (heavy goods vehicle) kilometres (CIAA, 2008).

Bibliography

British Retail Consortium (BRC) (2014), *A Better Retailing Climate: Driving Resource Efficiency*, www.brc.org.uk/retailingclimate.

British Sugar (2010), *2009-10 Corporate Sustainability Report*, Peterborough, British Sugar, http://britishsugar2.tinfishcreative.co.uk/Files/British-Sugar-Sustainability-Report-NAVIGABLE_2(2).aspx.

Campden BRI (2011), "Scientific and Technical Needs of the Food and Drink Industry", Campden BRI, Chipping Campden, Gloucestershire, www.campden.co.uk.

Carbon Trust (2015), "Energy efficiency and carbon saving advice for the food and drink sector", www.carbontrust.com/resources/guides/sector-based-advice/food-and-drink/.

Carbon Trust (2012), Food Preparation and Catering, www.carbontrust.com/media/138492/j7895_ctv066_food_prep_and_catering_03.pdf.

Carbon Trust (2011), Heat Recovery, www.carbontrust.com/resources/guides/energy-efficiency/heat-recovery/www.carbontrust.com/resources/guides/energy-efficiency/heat-recovery.

Chartered Institution of Building Service Engineers(CIBSE) (2009), *Energy Efficiency in Commercial Kitchens*, Cibse, London, United Kingdom, TM50.

Confederation of the Food and Drink Industries of the EU (CIAA) (2008), *Managing Environmental Sustainability in the Food and Drink Industry*, Brussels.

Department for Environment, Food and Rural Affairs (DEFRA) (2013), *Food Statistics Pocketbook 2013*, London, United Kingdom.

DEFRA (2008), *Comparative Life Cycle Assessment of Food Commodities Procured for UK Consumption through a Diversity of Supply Chains*, research undertaken for Defra by AEA, ADAS, Ed Moorhouse, Paul Watkiss Associates, AHDBM, Marintek, Defra project FO0103, 2008.

DEFRA (2007), *Direct Energy Use in Agriculture: Opportunities for Reducing Fossil Fuel Inputs - Final Report*, randd.defra.gov.uk/Document.aspx?Document=AC0401_6343_FRP.pdf.

Edwards-Jones G, L. Milà i Canals, N. Hounsome, M. Truninger, G. Koerber, B. Hounsome, P. Cross, E. York, A. Hospido, K. Plassmann, I. Harris, R. Edwards, G. Day, A. Tomos, S. Cowell and D. Jones, (2008), Testing the assertion that "local food is best": the challenges of an evidence-based approach", *Trends in Food Science & Technology*, Vol.19, pp: 265-274.

Environment Agency (2010), National Packaging Waste Database.

European Commission (EC) (2015), "Study on Energy efficiency and Energy Saving Potential in Industry and on Possible Policy Mechanisms", Paper prepared by ICF Consulting Ltd. for the EC DG Energy.

Federated Farmers of New Zealand (2015), www.fedfarm.org.nz/publications/Papers/

FoodDrinkEurope (2012), *Environmental Sustainability Vision Towards 2030.*

Food and Drink Federation (FDF) (2010), "Our recipe for change".

Food and Agriculture Organization of the United Nations (FAO) (2011), *Energy Smart Food for People and Climate – Issue Paper*, Rome.

Fritzson A. and T. Berntsson (2006), "Energy efficiency in the slaughter and meat processing industry - opportunities for improvements in future energy markets", *Journal of Food Engineering*, Vol. 77.

Helgerud, H. and M. Sandbakk (2009), *Energy Efficiency in the Food and Drink Industry: The Road to Benchmarks of Excellence*, ECEEE Summer Study 2009, European Council for an Energy-Efficient Economy.

James, S. and C. James (2011), Improving Energy Efficiency within the Food Cold-Chain, ICEF11 International congress on Engineering and Food, Athens, Greece.

Kaye, L. (2013), "Waste Heat Recovery: Getting Something Back", TriplePundit.

Kellogg (2014), *Kellog's Corporate Responsibility Report 2014.*

Lansink, A., E. van Ierland and G. Best (2003), *Economics of Sustainable Energy in Agriculture: issues and Scope*, Chapter 1 in van Ierland, E. and A. Lansink (eds.), *Economics of Sustainable Energy in Agriculture.*

Mancini, L. (2011), Food habits and environmental impact: an assessment of the natural resource demand in three agri-food systems", Universitá Politecnica delle Marche, Ancona, Italy, 2011.

Monforti-Ferrario, F., J.-F. Dallemand, I. Pinedo Pascua, V. Motola, M. Banja, N. Scarlat, H. Medarac, L. Castellazzi, N. Labanca, P. Bertoldi, D. Pennington, M. Goralczyk, E.M. Schau, E. Saouter, S. Sala, B. Notarnicola, G. Tassielli and P. Renzulli (2015), *Energy use in the EU food sector: State of play and opportunities for improvement*, European Commission, Joint Research Centre, Brussels, Belgium.

Nitzov, B. (2011), "Energy in food – Issues Brief", *Atlantic Council*, www.acus.org/publication/energy-food.

Packaging Digest (2012), "Walmart highlights sustainability efforts", March 2012. www.packagingdigest.com/smart-packaging/walmart-highlights-sustainability-efforts.

Pimentel, D., S. Williamson, E. Courtney, A. Gonzalez-Pagan, C. Kontak and S. Mulkey (2008), "Reducing Energy Inputs in the US Food System", *Human Ecology*, Vol. 36, No.4.

Pretty, J., A. Ball, T. Lang and J. Morison (2005), "Farm Costs and Food Miles: An Assessment of the Full Cost of the UK Weekly Food Basket", *Food Policy*, Vol. 31, pp. 1-19.

Ramírez, C., M. Patel and K. Blok (2006), "How much energy to process one pound of meat? A comparison of energy use and specific energy consumption in the meat industry of four European countries", *Energy*, Vol. 31, pp. 2047-2063.

Saunders, C. and A. Barber (2008), "Carbon Footprints, Life Cycle Analysis, Food Miles: Global Trends and Market Issues", *Political Science*, Vol. 60, No. 1.

Sim, S., M. Barry, R. Clift and S.J. Cowell (2007), "The Relative Importance of Transport in Determining an Appropriate Sustainability Strategy for Food Sourcing", *Int J LCA* , Vol. 12, No. 6.

Tassou, S., Y. Ge, A. Hadawey and D. Marriott (2011), "Energy consumption and conservation in food retailing", *Applied Thermal Engineering*, Vol. 31, No. 2-3.

Wang, L. (2009), "Energy Efficiency and Management in Food Processing Facilities", CRC Press.

Weber, C. and H. Matthews (2008), "Food-miles and the relative climate impacts of food choices in the United States", *Environmental Science and Technology*, Vol. 42, pp. 3508-3513.

Williams, A., E. Audsley and D. Sandars (2006), *Final report to DEFRA on project IS0205: Determining the environmental burdens and resource use in the production of agricultural and horticultural commodities*, London

Annex 3A.1

Actions by FoodDrinkEurope to reduce emissions

FoodDrinkEurope, the European food and drinks industry trade association, Environmental Sustainability Report (2012) highlights many of the changes that have already been put place in recent years and proposed some changes that could improve energy efficiency and reduce future emissions, for example:

- The Federation of German Food and Drink Industries established the "Energy efficiency in the German Food and Drink Industry" programme. Participants developed individual corporate action plans to reduce energy consumption and improve energy efficiency. Since 2008, 200 of the food and drink companies that have participated show that organisational measures can reduce energy costs by up to 20% with minimal effort.
- The UK Food and Drink Federation (FDF) has committed to working collectively to tackle climate change by making an absolute reduction in carbon emissions of 20% by 2010 and 35% by 2020 against a 1990 baseline. By 2010 FDF members reduced their emissions by 25% compared with the 1990 baseline.
- The food and drink sector has signed Long-Term Agreements (LTAs) on energy efficiency between government and industry in some EU Member States such as the Netherlands and Finland.
- The European food and drink industry has put in place practices aiming to utilise 100% of their agricultural raw materials through biological treatment of manufacturing food waste, thus reducing dependence on fossil fuels.
- Kellogg's manufacturing plant in Manchester, United Kingdom, contains a 4.9 MWe (megawatt electrical) Combined Heat and Power (CHP) plant that supplies 85% of the plant's current steam demand and approximately 50% of electricity demand. The use of the CHP plant reduces CO_2 emissions by approximately 12% annually.
- Large numbers of individual and collective actions to reduce waste through optimal use of by-products (e.g. for animal feed, lubricants, pharmaceuticals, bio-energy). Examples include the use of coffee grounds and other waste to create energy by Kraft, Nestle and Unilever and the UK's Food and Drink Federation members' commitment to send zero food and packaging waste to landfill by recovery and recycling (90+% already achieved).
- Optimised packaging by various mechanisms including reductions in the thickness and weight of cans, glass and plastic containers. Kraft Foods no longer packages assortment products into corrugated boxes but has switched to reusable rigid packaging. The associated storage efficiencies allow the company to better store unfinished goods on-site, reducing the need to move product by road to off-site storage. The polypans have an operating life of between five to ten years and their suppliers ensure they are recycled at the end of their operating life.

Various initiatives to reduce the impact of transportation on emissions have promoted improved planning and switching to more efficient trucks and switches to lower impact modes such as shipping or barges where feasible.

Chapter 4

Cross-cutting measures to improve energy efficiency in the food chain

While agro-food chain businesses – e.g. farms, food processors, retailers, distributors – vary substantially in terms of their scale, capital availability, operations and technology used, all generally tend to rely on energy for machine operations, lighting, heating, ventilation and air conditioning (HVAC), water and waste management. This Chapter first discusses the role of consumers in improving energy use and efficiency of the food chain. It then considers a sample of energy efficiency measures which are generally applicable to all stages of the food chain – both relatively low-cost operational changes and more substantial measures which require further investment. It looks beyond energy conservation measures, to include waste and water reduction and waste-to-energy investments as potential "win-win" means to reduce food chain fossil fuel consumption and waste, in addition to GHG emissions.

The role of consumers in improving energy efficiency

Lifestyle choices are key determinants of energy use in food consumption

The household stage of the food chain entails the purchase, storage, handling, domestic preparation and waste disposal of food by consumers. This element inevitably leads to issues of lifestyle choice, including decisions on meat consumption and the recycling of food waste.

At the household level, quantitative evidence on energy efficiency is limited. For the United States, studies have found that the household storage and preparation sector is the most energy-intensive segment of the food chain. The energy requirement in this sector is about a third of the total energy consumed in the food chain (Canning et al., 2017; Heller and Keoleian, 2000; Canning et al., 2010). Canning et al. (2010) find that the demand for convenience by consumers, in particular by consuming more prepared foods and a larger amount of food outside the home, both of which use more energy – is the principal reason for the increased use of energy in the agro-food sector. The study estimates that energy use for refrigerators, freezers, cooking and dishwashers will increase by 12% between 2009 and 2030. Over the same period, the US population is expected to increase by 22%, suggesting increasing efficiency in household machinery. However, in this segment of the U.S. food system, there is predominant use of electricity, unlike other segments, where fossil energy resources dominate agricultural production and transportation.

At the household level, energy efficiency options include: i) optimal use of domestic appliances (e.g. electricity and other fuels used for cooking, cleaning and food storage); ii) reduce fuel consumption in personal transportation for food consumption; iii) make purchasing decisions for food preparation equipment cognisant of the energy embodied in manufacturing this equipment; and iv) the minimisation of domestic food waste.

The relative price of energy is the most important driver on improving energy efficiency on the demand side. Energy use per unit of production increases as the price of energy decreases relative to the price of capital and labour, and *vice versa*. Higher energy prices increase the wedge between energy friendly and energy intensive commodities and thereby shift consumption towards the former. A variety of other factors, however, affect the energy reduction potential on the demand side, including total population, per capita food availability, the commodity content of food availability and the energy intensity of food-system production technologies (Canning, et al., 2017).

Reducing food waste also offers high potential for energy savings. Consumer behaviour could be altered by promoting food-waste awareness in education and general social programmes. Innovations in packaging and various policy measures, such as improving labelling, could support behavioural change to reduce food waste.

Significant energy demand reductions for food supply could, in theory, be achieved by moving human diets away from animal products, which are, in general, more resource-intensive than plant-based products.[1] However, this would need to be socially acceptable. In fact the reverse trend is true, particularly in Asia, where middle and upper income classes are tending to move towards a more western diet with higher animal protein content per capita.

Overall, policies that encourage food labelling, stimulate dietary change, reduce obesity, and avoid food losses can help reduce the energy demand of the agri-food sector at relatively low cost. However, social acceptance of such policies could be a barrier to implementation.

Reducing wastage across the supply chain is a win-win opportunity

Food losses and waste are a major cause for energy loss in food supply: it is estimated that between a third and a half of all food produced worldwide (up to 2 billion tonnes of food) is either lost or wasted (FAO, 2011).[2]

The problem of food waste is not limited to developing countries – one study suggests that as much as 40-50% of the food that is ready for harvest in the United States is not consumed and that U.S. households waste an average of 14% of their food purchases (Jones, 2006). Kader (2005) estimates that over 30% of the fresh produce (fruit and vegetables) harvested in both developed and developing countries is lost, with the rate being highest (20%) in the retail, food service and consumer parts of the system, whereas in developing countries the rate is highest in the distribution system from farmers to retail (22%).

The study by the World Business Council for Sustainable Development (WBCSD) estimates that rebalancing consumption at retailer and consumer level in mid-high income countries could decrease demand for food by 10%, resulting in 8% savings along the food chain (WBCSD, 2015). In the United States, reducing food losses by just 15% would provide enough food to feed more than 25 million Americans every year at a time when one in six Americans lack a secure supply of food to their tables (Gunders, 2012). Experts point at mismatch between supply and demand, poor purchase planning or unconsumed cooked food as main causes of food waste.

The EU-28 produces around 88 million tonnes of food waste every year, at an estimated cost of EUR 143 billion. 70% of EU food waste is produced by consumers, retail and food service sectors, while 30% is generated by the processing and production sectors (Gassin, 2016).[3] In France, a study by the French Environment and Energy Management Agency (ADEME) found that ten million tonnes of food was estimated to be lost or wasted across the food chain in France per annum, with an economic value of EUR 16 billion and a GHG impact of over 15 million tonnes CO_2 equivalent (Vernier, 2016). In the United Kingdom, the amount of food wasted post-farm gate is estimated by the UK Waste and Resources Action Programme (WRAP) to be about 10 million tonnes per annum, the equivalent of EUR 20 billion per annum (Rogers, 2016). The majority of waste is reportedly generated both at manufacturing level and the consumer end of the supply chain.

Much food is highly perishable and waste occurs throughout the food chain: in agriculture mainly during harvest and storage; in manufacturing due to mismatches between purchases and demands; and at the retail level, products unsold by the sell-by date. Consumers in turn are responsible for substantial waste of perishable products, notably salad greens.

Waste is also generated beyond the farm gate through product deterioration during transportation and storage and handling by food retailers and food service companies. Firms invest in storage and handling facilities and technologies to reduce wastage – the spread of refrigeration and cold storage facilities is one indicator of this – as losses of this sort inevitably affect profitability. They also use specifically designed packaging to protect food from damage. There is a potential trade-off involved in such innovations in that not only are they likely to increase energy use in the system, but packaging, if not recycled, can itself generate another form of waste.

Reducing food waste offers enormous potential for improving resource efficiency and energy savings.[4] One estimate puts the energy embodied in wasted food in the United States as equivalent to 2% of total energy consumption – roughly the same percentage as agriculture consumes (Cuéllar and Webber, 2010). Mehlhart et al. (2016) report that the energy-savings potential from preventing food waste for the EU28 totals between 1.02 EJ (exajoule) and 2.34 EJ per year, representing 2% to 4% of the overall primary energy demand in the EU28. This results in reductions in GHG emissions of approximately 290 Mt of equivalent carbon dioxide. This unexploited potential primarily arises from lifestyle and consumer behaviour rather than lack of technology diffusion or new products entering the market.

Energy use from waste in the food chain can essentially be reduced through a combination of improved technological, legal and consumption efficiencies, as well as through an increase in re-use, recycling and use of energy conversion technologies at all stages of the chain.

Increasing productivity sustainably in the food system will require stakeholders to examine product life cycles and governments will need to evaluate what can be done to promote efficient energy use and minimise product waste. This is already beginning to happen (for example, food retailers in some OECD countries are beginning to reduce the amount of plastic packaging they use).

Various initiatives are underway to promote the recycling of packaging materials. Many of the supply side initiatives involve the creation of networks, platforms or partnerships with participation from industry and other stakeholders. Pressure from the general public to reduce environmental damage can also be an important element in the process.

Governments can assist through the use of conventional measures, such as funding research, education and providing demonstrations of green technologies, but they can also aid the process by modifying existing regulations (e.g. those concerning product standards, or the use of waste products in feeding livestock), in order to bring about increased efficiency in the use of energy. Governments can also facilitate the development of new uses for "waste" in the system (e.g. composting to produce soil- conditioning products or using waste to produce bioenergy). Demand-side measures such as green public procurement are also receiving increasing attention, with governments acknowledging that insufficiently developed markets are often the key constraint for eco-innovation.

OECD has extensively examined the issue of food waste, notably in the series of papers and discussion at the 8th meeting of the OECD Food-chain Analysis Network on Reducing Food Loss and Waste in Retail and Processing, 23-24 June 2016 (www.oecd.org/site/agrfcn/meetings/8th-oecd-food chain-analysis-network-meeting-june-2016.htm) and at the 4th meeting of the OECD Food-chain Analysis Network on Food Waste along the Supply Chain, 20-21 June 2013 (www.oecd.org/site/agrfcn/4thmeeting20-21june2013.htm). A paper by BIAC (2013) points out that for the industry, reducing food waste is part of another win-win opportunity – reduced costs and productivity for the industry and reduced impact (climate change, land-fill, etc.) for society.

Waste minimisation as revenue-generating resource[5]

While some losses are unavoidable, excessive waste occurs at every stage of the food production system. The waste of food is also the waste of all of the embodied energy and other resources which were used in its production. The disposal of this waste itself involves energy use, and poses major challenges for food businesses, both economic and logistical. Increasing waste disposal costs have dramatically increased food production costs in recent years (Wang, 2009).

Given the costs of waste disposal, waste is increasingly being recognised by industries as a revenue-generating resource. When they are unable to prevent waste at source, food and drink manufacturers, for example, frequently use by-products not only as food, but also animal feed, fertilisers, cosmetics, lubricants and pharmaceuticals, and, critically, energy.[6]

Promoting energy efficiency in Korea's greenhouse horticultural sector through waste heat recycling

Since 2015, Korea has categorised waste heat from power plants or other sources as a renewable energy that can receive public financial support, as using waste heat reduces the cost of greenhouse heating in the country by 80% when compared to diesel. The Ministry of Agriculture, Food and Rural Affairs (MAFRA) launched a public information programme to promote the technology, which included providing information on the website of Rural Development Administration (which is a related agency) that gave local governments and farmers the necessary information for them to use waste heat. This document, "Industrial Waste Heat Map and Models for Utilisation", included the geographical locations of heat sources, the appropriate sizes for the greenhouse area requiring heating, as well as the basic design of various heating models. This information was also distributed in the form of a guidebook to designated technology centres in cities and provinces. Using this information, a farmer, the local government, and a power plant can jointly prepare and submit a project concept paper to MAFRA, which in turn selects a suitable project model for each region and provides support in the form of a grant and/or a loan.

Solid waste from fruit and vegetable processing is frequently dried, pelletised, and sold as low-value animal feeds. However, as mentioned earlier, the production of animal feeds from discrete solid waste such as this consumes large amounts of energy (Wang, 2009). Moreover, in the case of feed and fertiliser production from slaughterhouse wastes in particular, EU requirements for the minimum amount of time (and heat) that meat should be rendered, in order to reduce the risk of disease transmission via the feeds, also impact energy use. In other instances, food safety regulations can forbid the reuse of food waste of animal origin, with the result that this is sent to landfill (Salminen and Rintala, 2002).

Farms also produce substantial amounts of waste. In the majority of cases, however, waste is not generated in sufficient quantities by individual farms to justify any treatment. Nevertheless, if a number of farms are willing and able to co-operate, sufficient amounts of waste could be accumulated for recycling or other constructive uses, such as waste to energy, in spite of the additional transportation costs involved (DEFRA, 2007).

Consumers in higher-income countries cause a substantial part of waste in the food system

Consumer behaviour is also an important determinant of food waste. Indeed, evidence suggests that a substantial amount of food wastage in industrialised countries is in fact generated by consumers themselves, rather than the food industry *per se* (FAO, 2011). According to the US Environmental Protection Agency (EPA), food is the single largest component of municipal solid waste reaching landfills and incinerators, while it is reported by the organisation WRAP that almost half of all food waste comes from households and more than 60% of this is avoidable (BIAC, 2013).

In the United Kingdom, for example, households account for approximately 50% of total food waste generated (WRAP, 2012). Likewise, in Switzerland, the government has estimated that a third of all food is wasted, 50% of which is wasted at consumer level and 30% during processing, with less waste occurring at production and retail levels. Government actions have focused on awareness-raising and stakeholder dialogue, resulting in the joint decision by government and retailers to amend the labelling of fresh food from "sell by" and "use by" dates to "best before", which has had considerable positive impact on consumer waste.

There are several reasons why consumer food waste can be substantial in higher income countries (Blandford, 2013). First, affluence allows consumers to make precautionary purchases of food – to accumulate stocks that may not be needed. Second, which can amplify the effects of the first one, is response to "sell-by" or "use-by" labelling. The date stamping of products is used by food suppliers and retailers to manage inventories and to ensure that products are moved as quickly as possible to consumption. Sell-by dates provide an indication of when to consume a product in order to benefit from its highest quality, but this does not mean that the product will be unsafe to eat after that date. Use-by labelling is a stronger indication that it may be unsafe to consume the product after the date specified. Finally, consumers may react to scale economies in making purchasing decisions – either because they are faced with discounted unit prices for the purchase of multiple units of food items, or because they choose to overbuy in order to economise on the amount of time they need to devote to shopping.

Adapting and/or changing consumer behaviour is essential to reduce this waste (BIAC, 2013). Consumer awareness campaigns are an increasingly popular and effective means used by food manufacturers and retailers to manage food waste by households (BIAC, 2013). Nevertheless, unlike production-oriented measures, there is relatively little evidence of either the costs or the effectiveness of interventions to influence consumer behaviour.

Consumers can be prompted to achieve greater efficiency in food consumption and to reduce the amount of food waste when food prices and consumer incomes change. A study of household food and drink waste in the United Kingdom, for example, estimates that this declined by 13% or over one million tonnes between 2006/7 and 2010 (WRAP, 2012). The report concludes that while changes in products, packaging and waste collection have contributed to this decline other significant factors have been changes in consumer purchasing behaviour, particularly in response to higher food prices. The reduction of food waste documented by the report translated into an annual saving of roughly GBP 2.5 billion in food expenditures by consumers.

Food waste conversion to energy is gaining importance

The conversion into energy of food waste – such as crop residues including bran, husk, and bagasse; and food processing wastes including nut shells, rice hulls, meat processing residues, cheese whey, oils, fats, and wastewater – is a growing practice within agro-food chain businesses, enabling significant savings for companies in some cases, both in terms of waste disposal costs and reductions in the amount of energy purchased. As such, the integration of energy conversion processes into food processing facilities, for example, is considered to play an important role in the achievement of economic profitability, in addition to environmental benefits (Wang, 2009).

However, the French retail and distribution sector technical association (PERIFEM), which has undertaken a number of studies on food waste valorisation initiatives in the French retail sector, concludes that the energy potential of the biowaste was quite low (Gillier, 2016). It was also noted that the market for biowaste processing is very poor, and the cost of sorting and treating biowaste in compliance with safety and other regulations – such as the French "Grenelle II" law, which requires major producers of waste that is mainly composed of biowaste to sort this at source – is very high. The changing regulatory landscape with regard to waste and the lack of coherence between environmental, hygiene and waste regulations were also considered to pose challenges to waste management.

Food waste can also be difficult to use as a fuel source due to the varying characteristics and properties of different waste streams; regulations governing the reuse of certain wastes of animal origin; and the fact that wastes have to be prepared, stored, and transported to the location of energy conversion before they can be used to generate power or heat all of which add to energy and other costs (EPA, 2007). Comprehensive technical and cost analyses are therefore necessary to ascertain the feasibility and economics of energy conversion processes for various food wastes (Wang, 2009). Although the selection of an appropriate process is dependent upon the characteristics of the waste, the quantity of waste available, the efficiency of the energy conversion process, energy demands in the market and economic feasibility in the majority of cases, it is the form of energy needed that determines which energy conversion process is selected (Wang, 2009).

Bioethanol and biodiesel, for example, have been commercially produced to replace gasoline and fossil diesel, respectively. Meanwhile, the liquid oil produced by pyrolysis can be suitable for use in diesel engines and gas turbines. Finally, hot gas for steam generation, syngas and biogas, which are produced by combustion, gasification and anaerobic digestion, can be suitable for use at the production location.

Encouraging energy efficiency of water use

Water is an important cross-cutting area where technological and design changes can enable both water and energy efficiency. In the agricultural sector, for example, irrigation is a particularly energy-demanding input, mainly due to the energy needs of water pumping. Irrigated wheat production requires about three times more energy to produce the same amount of grain as rain-fed wheat, for example (Pimentel, 2009).

Energy requirements are largely dictated by crop choice, water availability (i.e. pressure differences between groundwater and surface water sources), decisions on scheduling in surface, spray or drip systems. In many countries this demand is also driven by unpriced or inefficient water allocations that discourage efficient water use. Inappropriate water pricing could lead to a misallocation of water resources by encouraging the production of higher value and more-water demanding crops and intensifying non-authorised groundwater withdrawals, particularly in cases with imperfect or weak enforcement of well permit or other regulations. Hence an indirect means of reducing energy use is to rationalise demand and move towards full cost recovery of water supply (OECD, 2010).

Examples of technological changes that have the potential to substantially improve water use efficiency on farms include smart irrigation scheduling, advanced irrigation management, the optimisation of pump sizes, and switching from flood irrigation to drip irrigation. The optimisation of pump sizes to take peak and off-peak season water requirements into consideration have been found to enable up to 35% energy savings in irrigation systems, for example (Jiménez-Bello et al., 2010; Moreno et al., 2010).

The collection, distribution, and treatment of drinking water and wastewater for use within the agro-food chain (e.g. for irrigation, cooking, heating, cooling and washing) consume vast amounts of energy (NRDC, 2009). Indeed, energy use for water treatment is expected to *increase* in future as water becomes less available due to the effects of climate change, and as more stringent water-quality rules and improved disinfection technologies – such as ultraviolet treatment and ozonation – are put in place. Improving infrastructure of drinking-water is a critical means to save both water and energy. It should be noted, however, that water reuse for irrigation may raise issues of water quality and the cost of treatment could be prohibited. For example, in Israel, although the expansion of the plan to recycle effluent water (mainly from sewage and industrial wastewater) has made a significant contribution to water and sewage management in the country, the process has given rise to a number of environmental and health costs (OECD, 2012).

Water re-use to cut energy usage

At a Kellogg facility in Ohio in the United States, new technology has been installed that reuses the water needed to produce steam for the cooking appliances. The technology enables the water to be re-circulated up to 50 times and is reported to also deliver significant energy savings. This project and others helped to reduce the facility's water use by 18% per metric tonne of food produced in 2014 (Kellogg, 2014).

Across the food chain as a whole, the recycling of wastewater that is suitable for reuse could provide a low-energy source of water supply. This is particularly the case in areas where significant amounts of energy are required to import and distribute water, such as in Southern California in the United States. For example, farmers are reportedly digging deeper water wells, and several areas in the state are exploring plans to build desalination plants. Both measures generate significant increases in energy use (WorldWatch, 2013). In areas such as these, recycled water can be delivered to users – usually at less cost than non-recycled water – for non-potable uses such as crop irrigation or sluicing down farmyards. In Orange County, California, treatment technologies are even used to purify wastewater to bottled water quality, before it is allowed to percolate into the groundwater basin for later use as potable water. This system reportedly uses only about half the energy that would be required to transport that water from Northern California to Southern California (NRDC, 2009).

Improving water efficiency can occasionally cause *negative trade-offs* for energy inefficiency. One example is the reduction of industrial cooling tower "bleed-off" (water that is periodically drained from cooling tower basins to prevent the accumulation of solids), which saves water but can, if excessively reduced, allow the formation of scales, which reduces energy efficiency (Galitsky et al., 2005). Another example is the case of an agricultural district in Andalusia, Spain, where the traditional open-channel network irrigation systems were modernised to a more water-efficient, on-demand pressurised system. Although the amount of water withdrawn for irrigation to farms was substantially reduced, maintenance costs increased by as much as 400%. The new pressurised pumping systems actually had higher energy needs that the previous gravity-based systems (Rodrigues-Diaz et al., 2011).

India is often cited as an example of how making subsidised electricity available to smallholder farmers can boost irrigation and food production. One of the criticisms of the Indian experience is that it lacked a fully integrated approach to natural resources, with over-pumping for irrigation putting massive pressure on groundwater sources (OECD, 2016).

Subsidised electricity encourages high energy use in Indian agriculture

State subsidies for irrigation, in the form of energy subsidies, are a central factor explaining groundwater demand and overdraft in India's Punjab and Haryana states. These have encouraged over-exploitation of groundwater resources in north west India for irrigation. In Punjab, electricity subsidies to agriculture represented accounted for 7% of the state's budget. In Haryana, where the farming sector consumes 40% of electrical power, the state government decided to pursue and extend power subsidies to farmers. Free energy supply for irrigation purposes is also made at the expense of the maintenance of the electricity network, which frequently collapses. Farmers have thus installed automatic power switches to pump water as soon as power is available. This uncontrolled and unmetered irrigation system encourages excessive water extraction. The withdrawal of the policy to supply free electricity is politically challenging, but targeting the subsidy to small and marginal farmers, linking the subsidy to irrigation requirement, metering and incentivising lower consumption are some options which can be resorted to reduce water consumption and enhance energy use efficiency.

Source: OECD (2016).

Capitalising on renewable energy opportunities

Using local renewable energy resources along the entire agri-food chain could help to improve energy access, allay energy security concerns, diversify farm and food processing revenues, avoid disposal of waste products, reduce dependence on fossil fuels and greenhouse gas emissions, and help achieve sustainable development goals. Land used to produce food also receives solar, wind and possibly hydropower resources with potential for heat or electricity generation. In addition a certain share of available biomass resources including crop residues, animal wastes, food process wastes that all result from agri-food chains could be converted into bioenergy and used for heat and power generation, as well as transport fuels.

Detailed assessments of each technology, together with issues concerning integration into existing and future energy supply systems, sustainable development, costs and potentials, and supporting policies are discussed in detail in the IPCC report, *Renewable Energy and Climate Change Mitigation* (IPCC, 2011).

Australia: Private sector engagement in energy-efficiency through the Emissions Reduction Fund

The Emissions Reduction Fund is the centrepiece of the Australian government's policy suite to reduce emissions. The Fund operates alongside existing programmes that are already working to reduce Australia's emissions growth such as the Renewable Energy Target and energy efficiency standards on appliances, equipment and buildings. The Fund provides incentives for Australian businesses and households to adopt practices and technologies that reduce emissions. The Fund has three elements: crediting emissions reductions, purchasing emissions reductions, and safeguarding emissions reductions. Companies, local councils, state governments, land managers and farmers can reduce emissions, increase their productivity and lower their energy costs by undertaking a range of eligible activities, such as upgrading to energy efficient equipment, replacing outdated lighting or generating power from waste. These activities allow them to earn Australian carbon credit units by reducing emissions. These credits can be sold to the Australian government through a carbon abatement contract, or to other businesses seeking to offset their emissions.

Quantum Power Limited, an Australian biogas company, is participating in the Fund with its bioenergy project at an abattoir in New South Wales to convert effluent waste from one of the country's largest livestock production facilities into electricity. The project will install biogas generators with capacity up to 1MW (megawatt) to convert organic materials in waste processing at the abattoir to biogas. The biogas is then refined and used as fuel for a renewable power station. This is the first project in Australia where biogas will be used to generate electricity in a way that offsets grid-supplied electricity. The biogas project is expected to replace more than 95% of grid-supplied electricity for the livestock enterprise.

Source: www.environment.gov.au/climate-change/emissions-reduction-fund.

Biofuels are liquid and gaseous fuels produced from biomass for use in the transport sector (IEA, 2012). These may fall into two broad categories: first-generation biofuels ("conventional" biofuels), made from food feedstock such as maize, sugar beet, sugar cane, and oilseeds, and second- or third-generation (often referred to as "advanced") biofuels, made from lignocellulosic biomass or woody crops, agricultural residues or waste. Given their reliance upon food crops, there is concern that first-generation biofuels compete directly with food and feed production for resources such as land, labour and fertiliser, and are responsible for contributing to upwards pressure on crop prices (OECD, 2013). Indeed, even the cultivation of second-generation dedicated energy crops such as woody biomass may place additional demands on land, water and other natural resources, and may have detrimental effects on the environment (OECD, 2013).As the focus of this report is on private sector initiatives in energy efficiency, supported by waste-to-energy investments, only biofuels generated from agricultural wastes and residues and waste produced downstream in the agro-food chain are discussed in this report (see the section on Food Waste). While, due to the absence of well-developed and secure supply chains for feedstocks, the lack of commercially-proven innovative conversion technologies, and the absence of stable and supportive long-term regulatory frameworks, these forms of biofuel have in the past been considered to be severely under-exploited as a source of renewable energy relative to their potential (https://europeanclimate.org/wp-content/uploads/2014/02/WASTED-final.pdf). In recent years, here have been developments in this area and a number of commercial-scale plants have been established in Brazil, the United States and Europe to convert bagasse, corn residues, wheat and rice straw into biofuels. Agricultural wastes and residues are also expected to become increasingly important as advanced biofuel feedstock in Asia (IEA, 2011).

Producing bioenergy from farming epitomises both the challenging trade-offs between different resources, and the positive synergies. While biofuel support may have aimed to help meet clean transport fuel targets in North America and Europe, much of the public debate has focused on negative impacts – such as diverting food crops like maize to biofuel markets, and competing with food crops for essential inputs – potentially contributing to food shortages and spiking food prices (FAO/OECD, 2011 – price variability). Yet advocates of bioenergy and of introducing an Integrated Food-Energy System (IFES) in general emphasise that many types and models of bioenergy can be efficient, sustainable and provide genuine development opportunities (Bogdanski et al., 2010). These require investments and that currently they cost more than conventional biofuels, but could be promoted in the framework of the implementation of the Paris agreement to help mitigate GHG emissions from the transport sector.

Biogas in China: Using waste to power homes, farming and agro-industry

The People's Republic of China has supported the installation of household biogas plants since the 1970s, with a major drive in government investment from 2003 onwards. This technology allows households to convert manure into clean cooking fuel ("biogas") and organic fertilizer. The programme has reached 100 million people, supplying a quarter of rural households. However, there are questions about the sustainability of the current design model. Fewer animals are kept as livestock in rural households today, meaning there is less manure to feed digesters, and migration to the city means there is less labour available to operate them. Other difficulties include inadequate maintenance and repair services and mixed results on the level of savings accruing to households. One institutional innovation that could reduce the burden on smallholders to raise livestock themselves is the 'District Biogas Farm' model piloted in Hainan. Small-scale farmers pay a larger district farm to rear pigs, and in return receive biogas and slurry at a discounted price, as well as revenue from the sale of the pigs.

Source: Best, S. (2014), "Growing Power: Exploring energy needs in smallholder agriculture", Discussion Paper, International Institute for Environment and Development, http://pubs.iied.org/pdfs/16562IIED.pdf

Careful policy assessment before implementation is necessary. Integration of energy and food production from biomass crops is technically feasible in many situations but needs to be managed carefully and in a sustainable manner. Detailed analysis on the sustainability of biomass use is being undertaken by such organizations as FAO, IEA, the Roundtable on Sustainable Biofuels and the Global Bioenergy Partnership.

Operational changes can lead to significant energy savings

Real energy efficiency gains are often perceived by companies as only possible through high-cost technological investments (OECD, 2015b). However, estimates from several studies – both of the agro-food sector and other industries – indicate that on average, energy savings of 20-30% can be achieved using operational changes alone, such as procedural and behavioural changes, without capital-intensive investment (Wang, 2009; OECD, 2015b). Whereas the potential for energy savings naturally varies according a number of factors, including the size and nature of the business, type of equipment used and local infrastructure, there is considerable evidence that operational changes can enable reductions in energy use across the entire food chain.

Examples of operational changes include the elimination of unnecessary energy consumption by equipment by switching off machinery when not necessary; the reliance upon daylighting whenever possible; proper insulation of process heating equipment; the rapid repair of water, steam, and compressed air leaks; prevention of drafts from badly-fitting seals; regular maintenance of energy-consuming equipment, etc.[7]

Compressed air, for example, is the most expensive form of energy used in an industrial plant because of its poor efficiency, which is typically about 10% from start to end use (Masanet, 2012). A production audit of a large Brazilian baked goods company revealed that the plant could reduce its electricity use by 350 MWh/year through the reduction of leaks in its compressed air systems. At an investment cost of USD 22 000, this would save USD 33 000 per year and pay for itself in nine months (International Finance Corporation, 2012 in ibid.). Motor maintenance programmes, meanwhile, can generate savings of 2%-30% of total motor system energy use. An IAC case study of a US dairy plant, for example, found that the implementation of a motor maintenance plan enabled savings which resulted in a payback period of about four months (Masanet, 2012).

Identifying energy savings through a factory energy auditing

A US bread factory with annual energy costs of approximately USD 820 000 per year conducted an energy audit that identified a number of cost-effective energy efficiency opportunities that could reduce its annual energy bill by 7%. The opportunities included improving motor controls, reducing compressed air usage, and installing occupancy sensors. The total investment costs of approximately USD 56 000 were expected to be paid back in one year through energy savings alone. An additional energy audit revealed that electricity use could be reduced by 25 MWh/year at another of the same company's facilities through the installation of low-cost occupancy sensors. At a cost of USD 2 900, the occupancy sensors would reportedly save USD 3 400 per year, thus paying for themselves in 10 months (BASE, 2012).

Cross-cutting technological or design changes are also possible

Beyond relatively low-cost operational changes, a number of more capital-intensive measures can be applied across the food chain, providing that businesses have access to the necessary finances. Examples include: i) measures to improve lighting efficiency include the replacement of lights and fixtures or the installation of innovative lighting systems; and ii) measures to improve heating, ventilation and air conditioning efficiency.

Post-harvest processes such as storage, drying, ventilation and cooling often have a great impact on energy use in the food chain. Drying is a typical method of preserving the quality of different agricultural products. It is a highly energy intensive operation since typically large quantities of water must be evaporated, due to the high moisture content of the harvested products, by heating, ventilating or a combination of both. The specific energy consumption for drying varies considerably depending on the type of the drying process, the scale of the dryer, the product to be dried, the initial moisture content, the meteorological conditions, and the age of the equipment. Cooling is an important process for the pre-treatment and post-treatment of agricultural products before and after preservation. Especially high value crops, like perishable fruit and vegetables, often have high requirements of post-harvest treatment, which are often very energy intensive.

Energy-efficient facility design

Mission Foods, a major tortilla manufacturer, designed its new production facility in Rancho Cucamonga in the **United States** to be as energy efficient as possible. Energy-efficient heating, ventilation and air conditioning and lighting technologies were installed, in addition to room occupancy sensors that turned lights off automatically, low-emissivity windows that reduced building heat gain, skylights that provided natural lighting, and new refrigeration system measures. The design changes enabled the company to reduce the electricity consumption of its new facility by roughly 18% compared with existing facilities, leading to annual energy savings of over USD 300 000 per year (Masanet et al., 2012).

Notes

1. Arguments for dietary change away from red meat products, and to reduce obesity, have been promulgated to reduce GHG emissions from the agri-food sector. GHG intensities vary widely with different food groups, with red meat, on average, being around 150% higher (in terms of CO_2-eq/kg) than chicken or fish.

2. At present, there is no commonly agreed definition of food waste or loss. FAO has attempted to develop "reference" definitions which define "food waste" as food which completes the food supply chain up to a final product, and is of good quality and fit for consumption, but which is nevertheless discarded and not consumed, and "food loss" as food, which during the food supply chain process, is spilled, spoilt or otherwise lost, or incurs reduction of quality and value before it reaches its final production stage (OECD, 2014). In the absence of a consensus on the definitions of food waste and food loss, however, this paper uses the term food waste.

3. The Commission is taking an increasingly active role in the area of food waste – food waste prevention is a component of the 2015 EU Circular Economy Action Plan and is addressed in the related EU Waste Framework Directive proposal, for example.

4. On the positive side, food waste also has the potential to play a role in renewable energy production as a feedstock for bioenergy production.

5. For packaging waste minimisation, please refer to the section on packaging.

6. Substances resulting from a production process which is not primarily aimed at producing such substances are defined by the food and drink industry as by-products and are not classified as waste (FoodDrinkEurope, 2012).

7. Energy auditing is the tracking of energy consumption and costs throughout a facility and the identification of opportunities to reduce energy use.

Bibliography

Berners-Lee, M., C. Hoolohan, H. Cammack and C. Hewitt (2012), "The relative greenhouse gas impacts of realistic dietary choices", *Energy Policy.*

Business and Industry Advisory Committee to the OECD (BIAC) (2013), Perspectives on Private Sector Solutions to Food Waste and Loss", paper presented to the OECD Conference, "Food Waste along the Supply Chain", 20-21 June 2013, Paris, www.oecd.org/site/agrfcn/BIAC%20Perspectives%20on%20Private%20Sector%20Solutions%20to%20Food%20Waste%20and%20Loss.pdf

BASE (2012), *Energy Savings for a Bread Plant*, Base Company, http://baseco.com/casestudies/Bread%20Plant.pdf.

Best, S. (2014), "Growing Power: Exploring energy needs in smallholder agriculture", Discussion Paper, International Institute for Environment and Development, http://pubs.iied.org/pdfs/16562IIED.pdf.

Blandford, D. (2013), "Green Growth in the Agro-food Chain: What Role for the Private Sector?", background paper for the OECD/BIAC Workshop, Green Growth in the Agro-food Chain: What Role for the Private Sector?, 24 April 2013, www.oecd.org/tad/sustainable-agriculture/Session%201%20BLANDFORD_FORM.pdf.

Bogdanski, A., O. Dubois, C. Jaimieson and R. Krell (2010), *Making Integrated Food-Energy Systems work for People and Climate*, FAO, Rome, Italy.

Canning, P., S. Rehkamp, A. Waters and H. Etemadnia (2017), "The Role of Fossil Fuels in the U.S. Food System and the American Diet", Economic Research Service, United States Department of Agriculture, ERS Report No. 224).

Canning, P., A. Charles, S. Huang, R. Polenske and A. Waters (2010), *Energy Use in the U.S. Food System* Economic Research Service, United States Department of Agriculture, ERS Report No. 94.

DEFRA (2007), *Direct Energy Use in Agriculture: Opportunities for Reducing Fossil Fuel Inputs - Final Report*, London.

Energy Design Resources (EDR) (2012), "Case Study: Tortilla Manufacturing Produces Energy-Saving Opportunities 2005".

FoodDrinkEurope (2012), *Environmental Sustainability Vision Towards 2030.*

Food and Agriculture Organization of the United Nations (FAO) (2011), *Energy Smart Food for People and Climate – Issue Paper*, Rome.

Galitsky, C., E. Worrell, A. Radspieler, P. Healy and S. Zechiel (2005), *BEST Winery Guidebook: Benchmarking and Energy and Water Savings Tool for the Wine Industry*, Berkeley, CA: Lawrence Berkeley National Laboratory, LBNL-3184.

Gassin, A.L. (2016), "Food waste prevention in Circular Economy Action Plan", Presentation to the 8th meeting of the OECD Food-chain Analysis Network on Reducing Food Loss and Waste in Retail and Processing, 23-24 June 2016, Paris, www.oecd.org/site/agrfcn/meetings/Introductory_AL%20Gassin.pdf.

Gillier, S. (2016), "Seeking ways for food waste valorisation in the French retail sector", French technical association for retail and trade (PERIFEM), presentation at the 8th meeting of the OECD Food-chain Analysis Network on Reducing Food Loss and Waste in Retail and Processing, 23-24 June 2016, Paris, www.oecd.org/site/agrfcn/meetings/Session%205_Sophie%20Gillier.pdf

Cuéllar, A. and M. Webber (2010), "Wasted Food, Wasted Energy: The Embedded Energy in Food Waste in the United States", *Environmental Science & Technology*, Vol. 44, No. 16.

Heller, M. and G. Keoleian (2000), "Life Cycle-Based Sustainability Indicators for Assessment of the U.S.", *Food System*, The Center for Sustainable Systems, Michigan, United States.

International Energy Agency (IEA) (2014), Capturing the Multiple Benefits of Energy Efficiency, IEA, Paris.

IEA (2012), *Technology Roadmap: Bioenergy for Heat and Power*, IEA, Paris.

IEA (2011), *Technology Roadmap: Biofuels for Transport*, IEA, Paris.

Intergovernmental Panel on Climate Change (IPCC) (2011), *Renewable Energy and Climate Change Mitigation.*

Jiménez-Bello, M.A., F. Martínez Alzamora, V. Bou Soler and H. Ayala (2010), "Methodology for grouping intakes of pressurised irrigation networks into sectors to minimise energy consumption", *Biosystems Engineering*, Vol. 105, No. 4, pp. 429-438.

Jones, T. (2006), *Using Contemporary Archaeology and Applied Anthropology to Understand Food Loss in the American Food System*, PhD, Bureau of Applied Research in Anthropology University of Arizona.

Kader, A. (2005), "Increasing food availability by reducing postharvest losses of fresh produce", *Acta Horticulture*, Vol. 682, pp. 2169-2176.

Mehlhart, G., I. Bakas, M. Herczeg. P. Strosser, C. Rynikiewicz, A. Agenais, T. Bergmann, M. Mottschall, A. Köhler, F. Antony, V. Bilsen, S. Greeven, P. Debergh and D. Hay (2016), *Study on the Energy Saving Potential of Increasing Resource Efficiency – Final Report*, Study prepared for the European Commission, Directorate General Environment, Brussels..

Masanet, E., P. Therkelsen and E. Worrell (2012), *Energy Efficiency Improvement and Cost Saving Opportunities for the Baking Industry*, Ernest Orlando Lawrence Berkeley National Laboratory, www.energystar.gov/sites/default/files/buildings/tools/Baking_Guide.pdf.

Moreno, M., P. Planells, J. Córcoles, J. Tarjuelo and J. Ortega (2010), "Energy efficiency of pressurised irrigation networks managed on-demand and under a rotation schedule", *Biosystems Engineering*, Vol. 107, No. 1, pp. 349-363.

OECD (2016), "Water Risk Hotspots for Agriculture", COM/TAD/CA/ENV/EPOC(2016)4/REV1, OECD Publishing, Paris.

OECD (2013), *Policy Instruments to Support Green Growth in Agriculture; A Synthesis of Country Experiences,* OECD Publishing, Paris, http://dx.doi.org/10.1787/9789264203525-en.

OECD (2012), *Water Quality and Agriculture: Meeting the Policy Challenge*, OECD Publishing, Paris. http://dx.doi.org/10.1787/9789264168060-en.

OECD (2010), *Sustainable Management of Water Resources in Agriculture*, OECD Publishing, Paris. DOI: http://dx.doi.org/10.1787/9789264083578-en.

Nestlé (2015), "Minister for Department of Energy and Climate Change sees how sweet waste is making sweet news for the environment", February 2015, www.nestle.co.uk/media/pressreleases/DECC-Minister-opened-AD-Plant-Fawdon.

Nestlé (2014), "Recycling coffee grounds as fuel", March, 2014, www.nestle.com/csv/case-studies/AllCaseStudies/Recycling-coffee-grounds-fuel.

Pimentel, D. (2009), "Reducing energy inputs in the agricultural production system", http://monthlyreview.org/2009/07/01/reducing-energy-inputs-in-the-agricultural-production-system.

Rodríguez-Díaz, J.A., L. Pérez-Urrestarazu, E. Camacho-Poyato and P. Montesinos (2011), "The paradox of irrigation scheme modernization: more efficient water use linked to higher energy demand", *Spanish Journal of Agricultural Research*, Vol. 9, No. 4.

Rogers, D. (2016), "Measurement of business food waste", presentation to the 8th meeting of the OECD Food-chain Analysis Network on Reducing Food Loss and Waste in Retail and Processing, 23-24 June 2016, Paris, www.oecd.org/site/agrfcn/meetings/Session%202_David%20RogersWRAP.pdf.

Salminen, E. and J. Rintala (2002), "Anaerobic digestion of organic solid poultry slaughterhouse waste – a review", *Bioresource Technology*, Vol. 83, pp. 13-26.

U.S. Environmental Protection Agency (EPA) (2007), "Biomass Combined Heat and Power Catalog of Technologies", U.S. EPA, Washington, D.C.

Vernier, A. (2016), "Food loss and waste: stock-taking and management along the food chain", presentation to the 8th meeting of the OECD Food-chain Analysis Network on Reducing Food Loss and Waste in Retail and Processing, 23-24 June 2016, Paris, www.oecd.org/site/agrfcn/meetings/Session%202_Antoine%20Vernier.pdf.

Williams, A., E. Audsley and D. Sandars (2006), *Determining the environmental burdens and resource use in the production of agricultural and horticultural commodities*, Final report to DEFRA on project IS0205, London.

World Business Council for Sustainable Development (WBCSD) (2015), Co-optimizing Solutions: Water and Energy for Food, Feed and Fiber, www.wbcsd.org/Projects/Climate-Smart-Agriculture/Resources/Co-optimizing-Solutions-water-and-energy-for-food-feed-and-fiber.

WorldWatch (2013), "Water efficiency key to saving energy, expert says", www.worldwatch.org/node/6007.

Waste Resources Action Programme (WRAP) (2012), *Courtauld Commitment 2 Voluntary agreement 2010 – 2012 Signatory case studies and quotes*.

Chapter 5

Unleashing the energy efficiency potential of the food chain

Notwithstanding progress, substantial potential for improving energy efficiency exists across the food chain. The existence of multiple barriers – policy, structural, behavioural and funding – discourages the private sector, including farmers, from making the best economic decisions. This chapter discusses these barriers and examines policies which could improve the energy efficiency of the food chain. Selected examples of country initiatives are also presented. Policies that seek to improve energy efficiency must also consider the synergies and trade-offs with policies addressing such issues as productivity, water use, and health and food safety.

The statistical data for Israel are supplied by and under the responsibility of the relevant Israeli authorities. The use of such data by the OECD is without prejudice to the status of the Golan Heights, East Jerusalem and Israeli settlements in the West Bank under the terms of international law.

Are we putting enough energy into improving energy efficiency in the food chain?

While the private sector has embraced energy-efficiency management practices and technologies, several analysts argue that there is a large *energy efficiency gap* or *energy efficiency paradox* – a wedge between the cost-minimising level of energy efficiency and the level actually realised – and that there are significant improvements to be made across the economy in general – and in the food chain sector, in particular (McKinsey & Company, 2010).[1] IEA energy projections to 2035, for example, show that as much as two-thirds of energy efficiency potential will remain untapped unless government policy changes (IEA, 2014). In the agro-food sector, evidence presented in the previous chapters shows that the energy efficiency of many food processing plants is low, when benchmarked against the best available technologies.[2]. However, precise estimations of the size and nature of the energy efficiency gap of the food chain have not yet been calculated.

There are numerous, well-developed management practices and technologies which can be used to reduce the amount of energy currently used in food production (Table A5.1). In many cases, the implementation of these approaches is a "win-win" situation for the producer, as the energy cost reductions achieved through implementation quickly repay the capital investment that is required. In the majority of cases, a payback time of five years or less can be achieved and, for a significant proportion of the technologies, the payback is three years or less (Gołaszewski et al., 2012, BASE, 2012; DEFRA, 2010). Moreover, important efficiency gains can be achieved without capital investment, by introducing procedural and behavioural changes.[3] Reducing food waste, for example, provides high potential for energy savings through resource efficiency and offers an important untapped energy-efficiency potential which policy could address (Mehlhart et al., 2016)

In agriculture, even though energy use is moderate compared with the other parts of the whole food supply chain, the Tassou et al. (2014) study for the United Kingdom contends that energy savings of up to 20% can be achieved through renewable energy generation and the use of more efficient technologies and "smart" control systems. In food processing, the authors argued that energy could be saved at the processing plant level by optimising and integrating processes and systems to reduce energy intensity (e.g. through better process control, advanced sensors and equipment for on-line measurement and intelligent adaptive control of key parameters). Likewise, they proposed the minimisation of waste through energy recovery and better use of by-products. In the food retail sector they note that significant progress in energy efficiency has been made in recent years, but that there still exists potential improvements in the efficiency of refrigeration systems, "heating, ventilation and air conditioning" (HVAC) and refrigeration system integration, heat recovery, and amplification using heat pumps, demand-side participation (DSP) system diagnostics, and local combined heat and power (CHP) systems and tri-generation. Tassou et al. (2014) also identify energy saving opportunities from the use of low-energy lighting systems, improved thermal insulation of the building fabric, integration of renewable energy sources and thermal energy storage systems. In catering facilities energy demand reduction can be achieved from the use of more efficient equipment, as well as via behavioural changes with respect to type of food consumed, food preparation practices, and environmental conditions in the premises.

The existence of multiple barriers discourages decision-makers, such as farmers, households and firms, from making the best economic choices. Many obstacles contribute to the limited uptake of energy-efficiency opportunities – such as subsidised pricing of energy, inadequate pricing of energy-use externalities, a shortage of financing, imperfect information, organisational inertia with respect to energy-efficiency investment opportunities by stakeholders in both the private and government sectors and systematic behavioural biases in consumer decision making (IEA, 2014; Gillingham and Palmer, 2013). Moreover, measuring the magnitude of profitable energy-efficiency investment opportunities that cannot be exploited, is by nature, a difficult procedure, and rigorous empirical research is lacking (Allcot and Greenstone, 2012).[4]

There are four broad groups of barriers that can be identified: structural, behavioural, availability and policy (Table 5.1). *Structural barriers* encompass issues such as limited know-how on implementing energy-efficiency measures, or fragmented and under-developed supply chains. Such barriers prevent an end-user from adopting an energy-efficient technology or practice: for example, low educational attainment and ageing farmers impede the adoption of new, potentially energy-efficient technologies.

Behavioural barriers include situations in which limited awareness or end-user inertia inhibit the pursuit of an opportunity. This may due to the lack of reliable information on costs and benefits, including energy prices that do not fully reflect all costs. As a result, the decision-maker is making a decision based on imperfect information. One example is a lack of awareness of food consumption efficiency behaviour; another is limited awareness of energy consumption differences, which might lead to the replacement of a broken water pump with the cheapest model available, rather than a more energy-efficient model with a lower total ownership cost.[5]

Availability barriers include situations in which the decision-maker is interested in and willing to pursue a measure, but cannot adequately access it: for example, a lack of access to capital might prevent an upgrade to a new heating system or the availability and diffusion of technology and innovations.

Policy barriers pertain to policy-induced market distortions which result in market conditions hindering energy efficiency. For example, energy subsidies can crowd-out public spending and private investment, encourage excessive energy consumption, reduce incentives for investment in renewable energy, and accelerate the depletion of natural resources. In fact, the low price of subsidised or otherwise non-market based oil prices may encourage a shift towards production which is more intensive in fossil fuels or energy usage more generally. In many countries, this runs counter to their broader environmental goals.

Table 5.1. Multiple challenges associated with pursuing energy efficiency in the food chain

Challenges	Strategies
Fundamental attributes of energy efficiency *Low-mind-share:* Improving energy efficiency is rarely the primary focus *Difficult to measure:* Evaluating, measuring and verifying savings are more difficult than measuring consumption, impairing investor confidence	**Opportunity-specific solution strategies** *Information and education* *Incentives and financing* *Third-party involvement*
Opportunity-specific barriers ***Structural*** *Human capital:* Ageing farmers, impeding adoption of potential energy-efficiency technologies *Insufficient knowledge:* Limited know-how on implementing energy efficiency measures *Fragmentation:* Fragmented and under-developed supply chains ***Policy*** *Pricing distortions:* Regulatory, tax or other distortions ***Behavioural*** *Risk and uncertainty:* Regarding ability to capture benefit of the investment *Lack of awareness/information:* About product efficiency and own consumption behaviour *Custom and habit*: Practices that prevent capture of potential ***Availability*** *Capital constraints:* Inability to finance initial outlay; limited access to capital for energy efficiency investments *Technology availability:* Insufficient supply or low diffusion of technologies *Installation and use:* Improperly installed and/or operated	**Components of an over-arching strategy** *Recognise energy efficiency as an important energy resource* Identify methods to *provide upfront funding* *Forge greater alignment* among stakeholders *Foster innovation adoption and development of new energy-efficient management practices and technologies*

Source: Adapted from McKinsey & Company (2010).

Eliminating inefficient fossil fuel subsidies can help to foster greater energy efficiency. OECD work has pointed to the removal of government support for fossil fuels as a potential "quick win". Government support to fossil fuel consumption and production in OECD countries and key emerging economies remains high, at USD 160-200 billion annually, according to a recent OECD report.[6]

Risk management attitudes and strategies can also be policy-induced barriers. For example, an unfavourable perception or treatment of risks could inflate the costs of energy-efficiency projects, or lead to the under-estimation of risks associated with changes in energy prices. Management of risks associated with energy costs and availability in agro-food businesses is largely determined by business size, with larger businesses being more likely to be proactive in managing risks from volatility in energy and commodity prices.

A policy environment that facilitates transmission of appropriate price signals to producers and consumers

Overcoming these barriers requires well-designed government policies to encourage energy efficiency and enhance competiveness (Porter and van der Linde, 1995; Ambec et al., 2011). The public sector has a vital role to play in providing the environment and infrastructure which would enable and promote action by the private sector. For example, government has an important role to play, in reforming fossil fuel subsidies or simply in helping to raise awareness of the costs and benefits of potential investment decisions. This would result in a win-win situation, with reduced costs for the private sector, combined with the social benefits from improved energy efficiency.

Market mechanisms must also be allowed to operate in order to provide incentives for the private sector to develop and adopt new technologies and to provide the financial resources to support green R&D. There is a role for public private partnerships in enhancing energy efficiency in the sector, particularly in the research and development of energy-efficient and energy-saving technologies.

The important role of the public sector in providing the regulatory environment to enable innovation and investment to thrive is recognised and encouraged by private-sector bodies:

> "The private sector has a crucial role to play in greening the agro-food chain, but it can only realise its full potential if appropriate business-enabling policies are put in place. The key drivers of green growth – namely investment and innovation – require the creation of enabling policy frameworks in which private sector-led and collaborative investment and innovation initiatives can thrive". (BIAC, 2013a)

> "To enable the retail sector to realise its ambitions in carbon reduction, it is vital that the regulatory framework provides for a clear and harmonised approach across business practices. Further simplification of a complex policy environment, together with a long term, clear and committed vision for low carbon in the United Kingdom, would give retailers the certainty and confidence required to make the significant investments needed to move to a low carbon economy." (BRC, 2014).

In addition, governments have a role to play in leading public opinion and setting an example, for example in the area of public procurement:

> "Government's core role in the United Kingdom food system is to correct market failures where they arise (for example distortions to the food economy caused by poor information, imperfect competition, the failure to price externalities and the under-provision of public goods). Government also has a role in setting the tone and direction of public debate about food, and a role in fostering cultural and behavioural change. This leadership and agenda-setting role can be a powerful complement to direct interventions." (DEFRA, *Food 2030 Strategy*).

Developing effective energy efficiency will require that economic, energy and natural resource issues be considered together, and that their mutual interactions and trade-offs be taken into consideration. Many of the commonly cited market failures are not unique to energy efficiency issues, and addressing them calls for a much broader policy response, such as an economy-wide price on GHGs to address climate change and comprehensive innovation policy to increase innovative effort. Conversely, information and behavioural failures – to the extent that they are substantial – tend to motivate more specific energy efficiency policies, provided that the benefits of the policies exceed costs.

The energy efficiency gap in the food chain has multiple explanations and the relative contribution of each differs between different types of producers and consumers. This heterogeneity in explanations poses challenges for policy makers, but also indicates when different policy interventions will most likely be cost-effective. Targeting policies towards market failures will improve cost-effectiveness.

A best-practice strategy should: i) identify the barriers to cost-effective efficiency investments and attempt to overcome them; ii) assess opportunities for energy efficiency improvements, and prioritise action in the agro-food chain and food consumers in which government policies are most likely to yield the largest, most cost-effective improvements; iii) set clear objectives and timelines, and establish evaluation methods; and iv) ensure coherence with energy, environmental/climate and economic strategies.

The array of policy solutions falls into two broad categories: i) measures to support informed choice; ii) measures to bring about changes to the market environment. Informed choice is the basis for consumer sovereignty which is integral to economic models of utility maximisation. Measures included in the former category are education programmes, labelling and social marketing (information from the state). Measures to change the market environment include food standards to regulate the nutrient content or sustainability of foods; taxes and subsidies on unhealthy foods or nutrients, or on GHG emissions; and regulation of the foods made available in schools or workplaces.

For both sets of measures the precise boundary between actions which provide a purely enabling environment for the private sector and actions that force the private sector (including consumers) to modify their behaviour, is blurred. A carbon tax could be considered as providing a clear framework for the private sector to innovate and adopt less carbon-intensive technologies, or it could be considered a highly interventionist and somewhat blunt instrument.

Policies tailored to addressing appropriate market failures can improve economic efficiency. If energy-use externalities (e.g. environmental or energy security externalities) are the only market failure, then taxes on energy use and emissions, or equivalent cap-trade programmes that internalise these externalities into energy prices are the most direct approaches to be used. These policies would unequivocally raise the price of energy, making energy-efficient products more financially attractive. If investment inefficiencies – consumers and firms do not undertake privately profitable investments in energy efficiency – are the only market failure, the first-best policy is to address the inefficiency directly – for example, by providing information to imperfectly informed consumers. If there are both investment inefficiencies and energy-use externalities, then taxes should be used in combination with some welfare-improving energy efficiency policy (Allcott and Greenstone, 2012; Gillingham and Palmer, 2013).

Appropriate pricing of externalities is a key to promote innovation and influence consumer behaviour. A clear, predictable carbon price is likely to be an important driver of change. Establishing an explicit price on greenhouse gas emissions through taxation or tradable permit systems is generally the most cost-effective means of creating strong economic incentives to reduce carbon emissions (OECD, 2013). Nevertheless, carbon pricing along the food chain could be politically challenging and is struggling to gain momentum in some countries. Once implemented, governments can also find it politically challenging to ensure that pricing mechanisms sufficiently reduce emissions (i.e. by increasing prices or significantly restricting the supply of permits in trading systems). Many energy-efficient and some low-carbon energy supply technologies are available today at zero or low additional net cost. Therefore to avoid locking in inefficient, carbon-intensive technologies, governments will need to intervene with targeted policies to bring down the cost of low-carbon alternatives and to create markets for technologies that are not yet fully commercial in the context of their energy efficiency strategies.

For most energy-efficiency opportunities a comprehensive approach will require multiple solutions to address the entire set of barriers facing cluster of potential energy gains. There is increasing agreement that the risk of the rebound effect associated with energy efficiency interventions can be reduced if energy-efficiency policy instruments, such as standards and regulations, are combined with other instruments, such as carbon taxes that increase the price of some energy products, or GHG caps that contain energy demand (Passey and MacGill, 2009). Table 5.2 provides a summary of potential barriers relating to energy efficiency, together with possible policy responses to address these problems in cases where they are found to be significant.

Table 5.2. Key barriers to energy efficiency and potential policy responses

	Barrier	Effect	Remedial policy tools
Awareness	Evaluating, measuring and monitoring energy efficiency is difficult	Opportunity not visible to decision-makers, impairing investor confidence	Measurement protocols and efficiency metrics; benchmarking; audits and reporting
	Low awareness of the value of efficiency	Energy efficiency is undervalued	Raising awareness and communication efforts; information and education
Policy	Policy-induced market distortions	Market conditions do not encourage efficiency	Removal of energy subsidies and other market distortions
	Energy market failures	Environmental externalities	Emissions pricing (tax, cap and trade)
	Unfavourable perception or treatment of risks	Financing cost of efficiency projects is inflated, or energy price risk is under-estimated	Better information on project and energy price risks, mechanisms to reduce efficiency project risk
Structural	Limited know-how on implementing energy-efficiency measures	Energy efficiency implementation is constrained	Capacity-building programmes
	Fragmented and under-developed supply chains.	Efficient opportunities are more limited and more difficult to implement	Programmes aimed at better market integration and overall economies.
			Forge greater alignment among stakeholders
Availability	Inability to finance initial outlays; liquidity constraints; limited access to capital for energy-efficiency investments	Under-investment in efficiency	Financing and loan programmes to stimulate capital supply for efficiency investments; support of new efficiency business; and financing models
	Technology availability and diffusion	Insufficient supply	Foster innovations and diffusion of technologies through R&D tax credits, public funding and incentives for early market adoption
		Low diffusion of technologies	
Behavioural	Lack of awareness and information about food-consumption efficiency and own consumption behaviour	Food waste	Education, information, product standards

Source: Adapted from OECD/IEA, *World Energy Outlook 2012*, Table 9.2.

A number of policies which could improve the energy efficiency of the agro-food sector are discussed below. Although they are discussed individually, it is probable that there are synergies between policies – for example, education and information measures linked to actions to promote product reformulation. This is important and it should always be remembered that a coherent policy mix to create an all-encompassing environment for green innovation is likely to be much more effective than the sum of the parts of a few *ad hoc* measures.

In general, energy-efficiency policies and programmes tend to be more effective if they are integrated into market strategies, addressing the range of barriers that are present in a particular locale (Nadel and Geller, 1996). In the US appliance market, for example, government-funded R&D helps to develop and commercialise new technologies; product labelling educates consumers (e.g. US Energy Star labelling); efficiency standards eliminate inefficient products from the marketplace; and incentives offered by some utilities and states encourage consumers to purchase products that are significantly more efficient than the minimum standards require. This combination of actions has led to dramatic and sustainable improvements in the efficiency of many types of appliances.

Forging competitive energy markets with appropriate regulation

Generally there are two approaches: fiscal measures (taxes and subsidies), which aim to influence the mix of food products society consumes (e.g. a carbon tax); and measures to influence the availability of foods or nutrients via the use of mandatory or voluntary standards.

Governments should periodically review regulations and subsidies to ensure that retail energy prices reflect the full costs of energy supply and delivery, including environmental costs. OECD and IEA analysis suggests that, globally, governments are spending a large sum of money (USD 500-600 billion) supporting fossil fuels, much larger than, for instance, what would be needed to meet the climate-finance objectives set by the international community, which call for mobilising USD 100 billion a year by 2020. This support is hampering global efforts to curb emissions and combat climate change. Recent OECD and IEA analysis indicates that phasing-out fossil fuel subsidies across the globe could lead to a 3% reduction in global GHGs in 2020, compared with business-as-usual.

Subsidies to fossil-fuel consumers often fail to meet their intended objectives of: alleviating energy poverty or promoting economic development and, instead, promote wasteful use of energy; contribute to price volatility by blurring market signals; encourage fuel smuggling and undermine the competitiveness of renewables and energy-efficient technologies. In addition, fossil fuel subsidies are often provided in conjunction with incentives that promote the use of renewable energies, thereby undermining policy coherence and sending confusing signals to producers.

Phasing out fossil-fuel subsidies would not only raise national revenues and reduce GHG emissions, but would also provide an impetus for investment, growth and jobs in renewable energy and energy efficiency. For this to succeed, well-targeted, transparent and time-bound programmes to assist poor households and energy workers who might be adversely affected in the short-term will be needed.

Using fiscal policies

Fiscal policies for encouraging the adoption of energy-efficient practices operate either through increasing the costs associated with energy use to stimulate energy efficiency, or by reducing the costs associated with energy efficiency investments. OECD (2013) discusses in some depth the effectiveness of the various policy instruments. Various forms of these measures have been implemented in numerous countries over the past three decades (see next section).

These include: energy or energy-related carbon dioxide (CO_2) taxes; taxes (and regulations) on the disposal of solid waste to encourage recycling and reduce the volume of waste going to landfill; grants and subsidies such as energy efficiency loans; use of deposit refunds on beverage containers and charging for plastic bags in supermarkets; or carbon trading schemes such as those that operate in the EU, New Zealand, Switzerland and California. In addition, integrated policies that combine a variety of financial incentives in a national-level energy or GHG emissions mitigation programme are also found in a number of countries. Such integrated policies are often national-level energy or GHG programmes that combine a number of tax and fiscal policies, along with other energy efficiency mechanisms, such as voluntary agreements. Mandatory standards have not been applied in the area of carbon footprint.

Energy taxes misaligned with environmental impacts of energy use

OECD analysis has consistently shown that price signals – as modified through energy taxes or emission trading schemes – are one of the best policy instruments to induce more sustainable patterns of energy use (OECD, 2001, 2006 and 2010). Regardless of their formal purpose, energy taxes influence the price and energy use patterns, and can provide incentives to seek alternative, cleaner technologies.

However, OECD analysis shows that energy taxes are poorly aligned with the negative side effects of energy use, and are having limited impact on efforts to reduce energy use, improve energy efficiency and drive a shift towards less harmful forms of energy (OECD, 2016). The low tax rates on many harmful forms of energy, and the existence of other measures that provide countering signals (e.g. fuel subsidies), suggest that countries, overall, do not harness the full power of energy taxes to reduce environmental harm in a cost-effective way.

For the food chain, energy or energy-related CO_2 taxes have been used in a number of countries to provide an incentive to improve the energy management at their facilities through both behavioural changes and investments in energy-efficient equipment. Taxes on energy or energy-related CO_2 emissions are now found in Denmark, Estonia, Finland, the Netherlands, Norway and the United Kingdom. In target-setting programmes that involve the use of energy taxes, such as in the Netherlands and the United Kingdom, rewards for meeting agreed-upon targets are provided in the form of a reduction of the required energy tax.

For agriculture, taxes on energy use are by far the single most important source of environmental tax revenue from the sector and are used in several OECD countries (OECD, 2014; Figure 5.1). However, tax rates on energy use for agriculture are much lower than those imposed on the whole economy (Figure 5.2). One of the reasons for this is that fuel used in agriculture is often exempt from tax. Fuel tax exemptions provide no signal with respect to external costs, thereby encouraging over-use. Given the rising awareness in of the negative side-effects of some sources of energy use and interest in investing in renewable energy sources, reconsidering the structure and level of taxes on energy can assist countries in pursuing their economic, social, and environmental objectives as effectively as possible.

Figure 5.1 Energy taxes in agriculture vary across countries

% in total environmental tax revenue from the sector, 2010 or most recent year

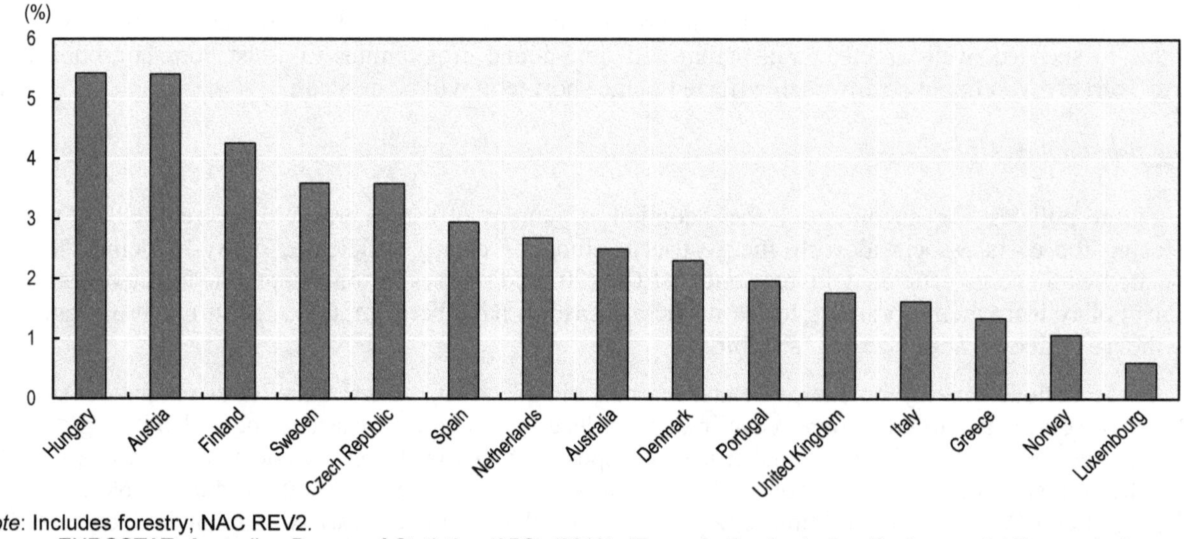

Note: Includes forestry; NAC REV2.

Source: EUROSTAT; Australian Bureau of Statistics (ABS) (2013), "Towards the Australian Environmental-Economic Accounts", *Information Paper*, Canberra.

Figure 5.2. Energy for agriculture is on average taxed less than in other sectors

Average effective tax rates on energy, EUR/GJ

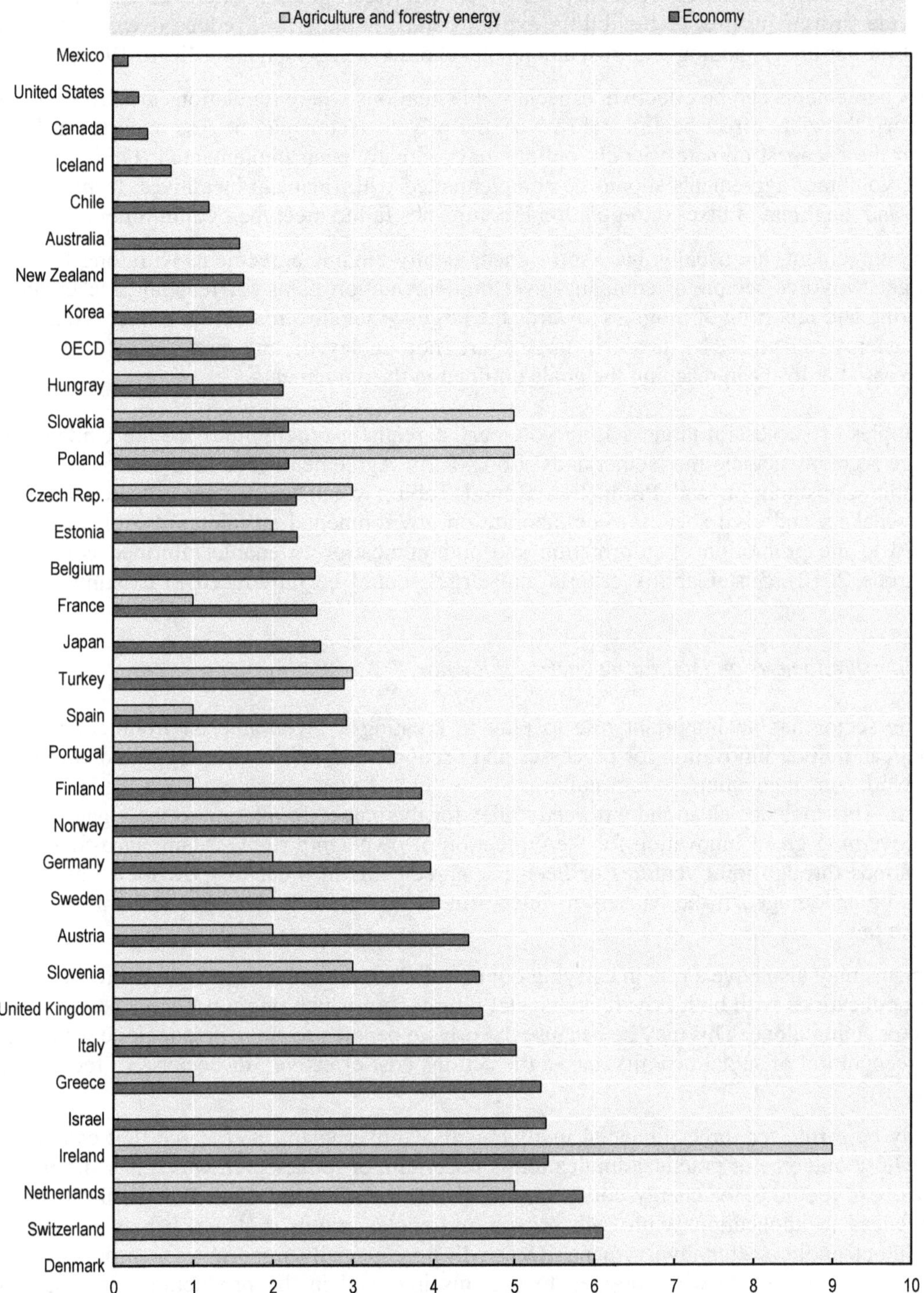

Note: Tax rates are as of 1 April 2013. Further details on the methodology can be found in the OECD (2015), *Taxing Energy Use 2015: OECD and Selected Partners Economies*, OECD Publishing, Paris, doi:10.1787/9789264232334-en.

Potential in voluntary agreements with the private sector

Compared to binding measures, voluntary agreements between governments and the private sector are argued to provide firms with greater flexibility, exploit company expertise, reduce overall costs, stimulate innovation and further understanding and trust among stakeholders (Segerson and Miceli, 1998).

Voluntary agreements can be effective, especially in situations where regulations are difficult to enact or enforce and when there are direct benefits. However, the private sector may be less open towards voluntary measures when the cheapest climate-friendly options have already been implemented (OECD, 2016). To be most effective, voluntary agreements should be complemented with financial incentives, technical assistance where needed, and the threat of taxes or regulation if companies fail to meet their commitments.

Voluntary agreements are usually: based on signed, legally-binding agreements with long-term (typically 5-10 year) targets; involve sector- or company-level implementation plans for reaching the targets; require annual monitoring and reporting of progress towards the targets; include a real threat of increased government regulation or energy-related GHG taxes if targets are not achieved; and provide effective supporting programmes to assist industry in reaching the goals outlined in the agreements.

Two examples of successful target-setting voluntary agreement programmes are the United Kingdom's Climate Change Agreements and the Netherlands' Long-Term Agreements (see next section). The European Food Sustainable Consumption and Production Round Table, which has gathered European food chain partners, policymakers and civil society to collaborate on environmental sustainability for the first time, is mainly oriented to the promotion of information and communication to enable informed consumer choice (FoodDrinkEurope, 2012). Sustainability criteria and targets could be introduced to existing reformulation platforms.

Innovation policy that focuses on facilitating energy efficiency

The public sector has an important role to play in creating a favourable environment for R&D and innovation that can deliver innovation for processes and products that are more energy (and carbon) efficient (see OECD work on innovation in agriculture, www.oecd.org/tad/agricultural-policies/innovation-food-agriculture.htm). This includes clear and enforced "rules for the game" in relation to the legal and regulatory environment governing green innovation for the protection of ownership rights. Many agro-food firms share process innovations through joint ventures or licensing agreements in order to grow their business and this propensity can be encouraged if the barriers to inter-firm co-operation in the development and use of new technologies are low.

Governments may also have a role in carrying out or collaborating with the private sector in research that may generate applications with both private and social returns, but which might not have been carried out by the private sector acting alone. This may be because the private benefits to firms acting alone were insufficient (whereas the recognition of social benefits makes the actions cost-effective), or because of too high a degree of uncertainty.

There may be a role for direct financial involvement of government in research that can contribute to greening but whose outcome or practical application is uncertain, or for research which has significant public good attributes (e.g. methods for energy conservation). There may also be a role for public involvement in R&D that will lead to applications with both private and social returns, but which may not otherwise be undertaken without public sector involvement. R&D will be especially important for progress in attaining energy efficiency in agriculture when applied to systems involved in the production process, operational activity and capital goods or farm infrastructure engaged in production.

Government-funded R&D contributed to the development and commercialisation of a number of new energy-efficient technologies in some countries and sectors. Experience has demonstrated that R&D can take many years to "pay-off", and that attention should be devoted to commercialisation and market development as well as technological advancement. Also, a prudent R&D portfolio includes high risk, potentially high pay-off projects, along with those involving lower-risk, incremental improvements.

Moreover, as innovative and cost-effective energy-saving technology is expected to continue to emerge, fostering innovation in the development and deployment of next-generation energy-efficiency technologies through R&D could ensure sustained productivity gains. R&D in this area can generate private benefits by leading to economies in the use of increasingly costly resources or controlling costs incurred in meeting environmental standards. It can also make products more attractive to consumers who are concerned about the environmental impact of their purchasing decisions, and thereby help to increase sales for a firm. Both of these developments can contribute to profitability and the survival of the firm in a competitive business environment.

Under the assumption that there are economic returns to innovation either through cost reduction or enhancement of sales, an important issue is whether mechanisms are in place to allow firms to obtain the necessary finance to fund the development of new technologies. Large and relatively wealthy firms are often able to fund their research and development activities from their earnings, but this is likely to be more challenging for small and medium enterprises. They will need to obtain external funding and depending on conditions in financial markets this may not be readily available. In addition, smaller firms may simply not be large enough to support investment in R&D of a sufficiently large scale, even if financing is available. Some collective action on the part of firms may be necessary to make such investment viable.

Enabling policies will continue to drive energy-efficiency investment, even in a low oil-price environment

Investing in energy efficiency may be risky due to the irreversibility of the investment and fluctuating energy prices. If energy prices fall, then the return on the investment falls as well. A number of studies report strong linkages between the prices of energy and non-energy commodities, using a variety of methodologies (Baffes et al., 2015). Energy prices were found to be the most important factor explaining movements in food prices for the post-2005 period (Baffes and Haniotis, 2016).[7]

This implies that increases in agricultural commodity prices will mean greater revenues for agricultural businesses, which could have the effects of improving the investment possibilities for the company. This could lead to investments in energy-efficient storage systems, machinery or management of land. Likewise, an increase in energy prices will translate directly into higher cost prices, especially for energy-dependent food sectors, stimulating the pursuit of greater efficiency in energy use.

Lower energy prices would benefit the agricultural sector primarily because of reduced production and transportation costs. But falling oil prices would, most likely, be accompanied by declining agricultural prices – and this would lead to the cessation of actions to promote energy efficiency. Baffes and Haniotis (2016) found that a 45% decline in oil prices could be expected to reduce agricultural commodity prices by about 10%.

Lower oil prices also impact on the food chain through several channels. Most importantly, falling fuel prices are expected to put downward pressure on other commodity prices, especially on natural gas, which, in turn, is a key input of fertiliser production (especially nitrogen-based production).[8] Lower oil prices could also reduce the opportunity cost of biofuel production. However, the declining attractiveness of biofuels production in an environment of low oil prices is likely to be mitigated by current policies. Because most of the diversion of food commodities to biofuels is policy-mandated, the increase in oil consumption triggered by low oil prices may, in fact, increase the diversion of grains and oilseeds to the production of biofuels.

Marshall et al. (2015) analysis suggests that in the United States the overall effects of lower energy prices in 2015`and 2016 on agriculture – acreage and prices – are anticipated to be modest because changes in production are small and crop acreage supply response is inelastic to changes in producer net returns (i.e. changes in producer net returns lead to relatively small acreage changes).[9] Effects on individual commodities reflect the importance of energy in production costs and whether the commodity is used as an input in the production of biofuel. Declining natural gas prices in the United States due to shale gas did not lower fertiliser prices, because the fertiliser price is determined globally, while natural gas prices are regional (Hitaji and Suttles, 2016). Hausman and Kellogg (2015) find that the U.S. ammonia price does not substantially diverge from global prices after 2007, despite the large decrease in the U.S. natural gas price.

In certain countries, the fall in oil prices has also provided favourable conditions for the reduction of end-use fossil fuel subsidies, which undermine the economic attractiveness of energy-efficiency investments. Several economies, including Austria, the Netherlands, India, Indonesia, Malaysia and Mexico and have recently either cut or abolished their fuel consumption subsidies.

Although recent pronounced downward shifts in global oil prices and regional gas prices have reduced the financial attractiveness of investing in energy efficiency, the IEA has found no evidence of this being the case for the economy as a whole (IEA, 2015). Energy-efficiency investments are set to keep growing, driven by more assertive and more comprehensive policies, such as tighter regulations on new buildings, products, vehicles and utilities, which recognise energy-efficiency measures as being among the most cost-effective means of helping to tackle energy security, productivity, local air pollution and climate-change challenges (IEA, 2015).[10]

Building public awareness of energy-saving opportunities

Increasing awareness on how and where energy is used is crucial in designing energy-efficiency measures. To establish whether large-scale investment in energy efficiency in order to reduce energy demand is cheaper than new energy supply, it is important to have reliable information on the net benefits of saving energy. This includes the cost of programmes, the value of the energy saved and any co-benefits. Until recently, energy efficiency has been accorded low priority by producers and consumers. For example, the priority of the food industry is the safety, quality and hygiene of the delivered product. Efficient-energy use is generally seen as a priority only when energy prices are high, rather than as a pre-requisite.

With better access to credible information about their energy-use and the options available to them, businesses will be able to make smarter energy decisions. A good level of information and analysis can reduce risk and uncertainty, and can help firms that are seeking to obtain finance for energy-efficiency projects, by endorsing the projects' credibility.

The energy-efficiency policy in OECD countries, in general, has focussed primarily on increasing the energy efficiency of buildings, appliances, vehicles and industrial operations. Less attention has been devoted to changing consumer behaviour (e.g. encouraging people to change their attitudes).

Public information campaigns exploit media communication and other social marketing tools to improve individual and social knowledge. This is by far the most common policy intervention in relation to healthy eating (Capacci et al., 2012). One possible reason for the popularity of information campaigns within the nutrition policy portfolio may be that they do not impose direct restrictions or direct costs upon industry. Campaigns promoting positive messages have been at the heart of the various "5-a-day" (or more in Denmark, Japan and Australia, for example) campaign promoting the consumption of fruit and vegetables.

It is important to remember that public awareness campaigns in relation to energy efficiency are relatively recent, so their long-term effectiveness is difficult to assess.[11] Very little is known, theoretically or empirically, about how social norms change over the long term. However, in relation to sustainability and the carbon footprints of foods, social marketing is an important component of any policy package. Governments must assess the evidence and give a clear message to consumers and industry as to what is important or unimportant and hence what changes in food consumption and production are to be encouraged. By this means, confusion – for example, as to whether imported food is better for the climate than home-produced food (i.e. the food miles debate) – can be clarified.

Sustaining policy impacts over the longer term requires sustained investment. Short-lived social marketing campaigns may be doomed to fail. For example, evaluation of a fruit and vegetables information campaign in Western Australia showed a significant increase in consumption during the campaign, but a falling-off immediately afterwards (Pollard et al., 2008). Substantially more research is needed on the causal links between campaigns and long-term patterns of behaviour and consumption.

Another argument in favour of social marketing is that it is widely accepted both by the public and industry. However, public information campaigns may work best when implemented in synergy with other measures, such as increased availability or improved labelling (EATWELL, 2012). Moreover, successfully boosting changes in consumer habits regarding diets and food preparation would be difficult, unless efforts to make this change are linked to achieving national health objectives. For example, establishing financial incentives or taxes that discourage people from eating foods with high levels of animal fat could be part of national efforts to reduce heart disease and obesity.

Labelling that informs consumers about the composition of foods and to help them make informed choices about what they eat could help overcoming informational market failures. Nutrition and eco-labelling aim to inform consumers about the composition of foods and enable them to make informed choices about what they eat; it is a necessary condition for overcoming informational market failures, but does not overcome externality or public good issues associated with food consumption. In principle though, educated and informed consumers can be appropriately targeted by food companies through their product development and marketing, enabling the market (for sustainable foods) to expand.

High participation and high savings success: Efficiency Vermont's Agriculture Programs, Wisconsin's Farm-Save Energy Program and Xcel Energy's Farm Energy Conservation Improvement Program

These three programmes are notable for their high participation and high energy-savings results, both evident through each programme's extensive evaluation process. The FarmSave Program is an education-only programme and the other two programmes include both education and financial incentives to the agriculture sector. The FarmSave Program is implemented through public benefits funding by the En-Save Energy consulting group. The outreach and education programme attempts to reach out to all farm types and has a broad array of experts to assist farmers with increasing their energy efficiency. Participants in this programme are also educated about possible financial incentives to perform upgrades, so in this way, the programme contributes to even more savings.

The Efficiency Vermont Programs also target all farm types, and working in tandem with the other programmes in Vermont has nearly a 100% contact rate with farmers in the state. The programme is a full-service one, including farm energy audits, and education, technical, and financial assistance to carry out recommendations from the audit.

The Xcel Program is also implemented by EnSave Energy, and is comprised of an education and outreach programme through both workshops and direct contact with farmers. The programme follows up on the education program by offering cash incentives to the participants to assist in the purchase and installation of energy-efficient equipment. Because of their focus on evaluation, these programmes are able to show that high participation often leads to high savings.

Source: Brown, Elliott and Nadel (2005) "Energy efficiency programs in agriculture: design, success, and lessons learned", *American Council for an Energy-Efficient Economy (ACEEE)*, Report No. IE051.

As well as encouraging consumers to change their behaviour, labelling encourages product and process innovations aimed at producing 'cleaner' products, showing lower energy consumption (GHG emissions); such measures may enhance brand reputation and the ability to demonstrate with confidence and credibility firms' commitment to lowering their environmental impact (PEF, 2014). The criteria set for the Dutch Choices logo resulted in significantly lowered levels of sodium in existing products, and higher fibre content of newly developed products and lowered levels of saturated fatty acids and added sugars in dairy products (Vyth et al., 2010).

A major challenge with introducing energy labelling is the large and growing number of "quality" logos already in existence and the consequent difficulty in attracting consumer interest, recognition and behavioural response to yet another logo or table of information (Hasler, 2008; Mhurchu and Gordon, 2007). As a consequence, Lang (2013) has proposed an omni-label indicating compliance with a range of quality criteria, but the practical problems of agreeing a fair system, communicating it and keeping it up-to-date would be formidable. Further research into understanding the various options and obstacles is necessary

Designing labels for retail food packaging that specify the amount of energy used in the production, processing, packaging and distribution of the product could encourage consumers to consider the energy and GHG implications when making their purchases. However, this is a complex undertaking and would require international standards for measuring energy consumption using standardised LCA methodologies to assess each stage of the food chain (Ziesemer, 2007).

Although information measures are generally the least controversial – at least with respect to diet and health – according to Capacci et al. (2012), they depend on consumers preparedness and, often, willingness to pay, to modify their diets and, in the face of a public good like climate change, face the usual public-good problem of under-production and consumption; only direct market intervention methods have the potential to overcome this issue fully. For example, the *US ENERGY STAR* labelling programme exemplifies the impact that a well-conceived, widely promoted labelling and education effort can have. But these "soft policies" tend to be more effective if combined with financial incentives, voluntary agreements, or regulations (Nadel and Geller, 1996).

Selected country examples

Over the last several decades, several governments have implemented policies to manage the demand for energy and improve energy efficiency along food systems. These policies tend to be set within a broader constellation of policies designed to reduce energy use for the overall economy and encourage more energy-efficient behaviour.

Australia: Increasing private sector investment through the Clean Energy Finance Corporation

The Clean Energy Finance Corporation (CEFC) is an Australian Government-owned bank that was established to facilitate financial flows into the clean energy sector. (www.cleanenergyfinancecorp.com.au/what-we-do.aspx). Its main function is to invest in emissions reduction activities in Australia. In particular, the CEFC's investment objectives are to increase investment in Australian-based projects related to energy efficiency, the commercialisation and deployment of renewable energy, and low-emissions technologies. This is done through an investment strategy focused on cleaner power solutions, including solar, wind and bioenergy; and a better built environment, with investments to drive more energy efficient property, vehicles, infrastructure and industry. The CEFC also invests with co-financiers to develop new sources of capital for the clean energy sector.

The CEFC applies a commercial approach when making investment decisions, focussing on projects and technologies at the later stages of development. In line with its policy intent, the Corporation considers the positive externalities and public policy outcomes when making investment decisions and when determining whether concessional funding should be granted.

The CEFC has partnered with two of Australia's largest banks to provide loans to energy efficiency projects. AUD 120 million has been provided to projects that will improve energy use through the National Australia Bank. AUD 100 million of business loans has been co-funded by the Commonwealth Bank and the Australian Government. Funding is for projects valued up to AUD 5 million, and is aimed at cutting energy and carbon emissions in small to medium enterprises, including farms. Some of the agro-food chain projects supported under the CEFC include the installation of solar panels at Australia's largest beef company, AACo; generating energy from waste at Darling Downs Fresh Eggs in Queensland; and a solar thermal system to desalinate sea water to grow tomatoes at Sundrop Farms in South Australia.

Denmark: Agriculture plays a key role as a green energy supplier in the transition to fossil fuel independence

A central element in Denmark's Green Growth Strategy is the emphasis placed on the development of renewable energy in the agricultural sector. The Green Growth Strategy is part of Denmark's long-term energy goal to become completely independent of fossil fuels use by 2050. In particular, the role of the agricultural sector as a supplier of energy is to be strengthened, with up to 15% of arable land to be used for energy crops – which represents a 16-fold increase in production of energy coming from agriculture – and the share of farm animal manure to be used for green energy is to be increased from 5% to 50% by 2020.

Policy initiatives to reach these targets include annual financial support for starting investments in biogas. The treatment of slurry for biogas is voluntary and farmers receive a premium of DK 75 per m^3. Up to 100 farmers can use one biogas installation. Under this scheme a grant covering up to 20% of the investment in the plant can be provided. The remaining funds are provided by a 60% loan that is guaranteed by the local municipality and 20% of own financing. Municipalities are obliged to include the construction of biogas plants in their municipal planning, as well as the allocation of grants for selling biogas to co-generation plants and the natural gas net.

European Union: Energy Efficiency Directive an important driver

The 2012 Energy Efficiency Directive (EED) establishes a set of binding measures to help the EU achieve its 20% energy efficiency target by 2020 (https://ec.europa.eu/energy/en/topics/energy-efficiency/energy-efficiency-directive). Under the Directive, all EU countries are required to use energy more efficiently at all stages of the energy chain from its production to its final consumption. It covers all sectors except transport. On 30 November 2016, the European Commission proposed an update to the EED, including a new 30% energy efficiency target for 2030.

The EED covers a wide variety of approaches and measures. These include indicative national energy efficiency targets, strategies to reduce energy consumption of existing buildings, the introduction of energy efficiency requirements in public procurement, obligations for energy companies to help customers save energy, and improvements to customer metering and billing. As a European Directive, the EED is a legal instrument that requires the transposition of its requirements into national laws by member states while also providing member states with a certain degree of flexibility in respect of implementation.

Finland: Multiple energy efficiency measures and support for biogas for promoting renewable energy

Finland's National Energy Efficiency Action Plan 2014 includes five energy efficiency measures related to the agricultural sector: investment in heating plants (including farm-scale biogas plants); investment in constructing fresh grain silos; investment in energy efficiency of cattle sheds and pig farms; farm re-parcelling projects; and the farm energy programme (https://ec.europa.eu/energy/sites/ener/files/documents/2014_neeap_en_finland.pdf).

Investment support ranges from 10 to 40% of the investment cost. The energy saving effect of these measures has been estimated to rise to a level of 17% by the year 2016 and to a level of 21% of total energy consumption of the agricultural sector by 2020.

The highest energy savings can be achieved by investing in heating plants that use biomass instead of oil. The purpose of the investment support is to promote: the increased use of renewable energy sources; the more efficient use of energy and energy saving; the adoption of new energy technologies; and the reduction of environmental damage from energy production and use.

For example, the requirement for granting the support for the construction, expansion or renovation of a boiler house is that the boiler house must use a source of renewable energy, including biomass. If peat is used as an energy source, the boiler house must also be able to produce heat from wood or another renewable energy source. The support is not granted for costs resulting from the utilisation of oil, hard coal or other similar fossils. The amount of support depends on the nature of the target receiving the funding. The support is granted only for investment where the production does not exceed the average annual energy consumption of the farm.

The Farm Energy Programme is a voluntary energy efficiency agreement signed in 2010 between the Finnish Ministry of Agriculture and Forestry and producer organisations for the period 2010-16. The programme encourages long-term improvements in energy efficiency and the use of renewable energy sources. It also aims to help meet national climate and energy targets that have been set to fulfil Finland's international obligations.

The programme is the main tool for achieving the energy-saving targets established by the EU in the agricultural sector and sets indicative target for improving energy efficiency of 9% of energy use of participating farms by 2016. The programme covers energy efficiency in electricity, heating, farm machinery fuels and water use.

The goal for farms participating in the programme was to represent at least 80% of the total energy consumption of the agricultural sector. In terms of implementation, depending on the scale of their energy use, farms joining the programme are provided with an InHouse Control Plan for small energy users and Farm Energy Plan and Farm Energy Audits for the largest energy users. A government subsidy, of up to 85% of eligible costs (from 2015 100 %), is available for preparing a Farm Energy Plan. The revision of the programme started in January 2015 following the introduction of the EU-financed rural development programme for the period from 2014 to 2020. The final evaluation report was published in October 2015; it was found that the goal had been overly optimistic, with only 484 farms joining the programme. On the other hand, the report found the participated farms found the programme and its services very useful.

Renewable energy

The Finnish Long Term Climate and Energy Strategy, launched in 2008, recommended that the use of biogas should increase by 2020.[12] In order to promote combined head and power production using biogas, a market-based feed-in tariff scheme (only for those biogas plants which produce at least 150 kVA of electric power) has been introduced, which is financed from the state budget.

Government funding is also available for relevant research, investigation, training and communications projects to promote the establishment of bio-energy production plants, as well as for pilot projects applying new research data and technologies. A particular objective of the support is to promote the construction of biogas plants in areas with large farm animal populations and consequent environmental impacts. Biogas plants can produce electricity, heat or transport fuels, but, in addition to producing renewable energy, they also have positive environmental impacts brought about by the improved use of manure and reduction of GHG emissions. The support is primarily targeted at biogas plants that are not accepted under the terms of the electricity feed-in tariff scheme.

In addition to the support for the production and use of biogas, bio-energy investments in rural micro-enterprises and small - and medium-sized enterprises received financing under the Finnish 2007-13 Rural Development Programme. For example, support was granted to bio-energy product refinement, energy production from biomass or other construction investments related to bio-energy business activities.

France: Focus on energy-intensive processes, the use of tractors and renewable energy

Improvement in the energy performance of agricultural holdings is one of the commitments of the *Grenelle Round Table,* set out in Article 31 of *Grenelle I* in 2007. Support measures focus mainly on more energy-intensive processes, in particular heated greenhouse productions, intensive productions, the use of tractors and on renewable energy (https://ec.europa.eu/energy/sites/ener/files/documents/2014_neeap_en_france.pdf). Key programmes include the Energy Saving Certification system (*Certificats d'économie d'énergie* – CEE) and the Agricultural Competitiveness and Adjustment Plan (*Plan de compétitivité et d'adaptation des exploitations agricoles – conciliation de la performance économique et écologique* – PCAE).

The Energy Saving Certification system

The Energy Saving Certification system (CEE), introduced in 2005, aims to create energy savings particularly in those sectors where energy savings can be more commonly found, such as buildings but also including small- and medium-sized industries, agriculture and transport. Under the CEE, energy suppliers (electricity, gas, heating oil, LPG, heat, refrigeration) must meet government-mandated targets for energy savings. Suppliers are free to select the actions to meet their objectives, such as informing customers how to reduce energy consumption, running promotional programmes, providing incentives to customers and so on. Those exceeding and undercutting their objectives can trade energy savings certificates as required for common compliance. Energy suppliers who do not meet their obligations must pay a penalty.

About forty major suppliers of electricity, gas, heat and refrigeration, as well as more than 2 000 distributors of heating oil and some forty consumers of automotive fuels are subject to savings obligations. The energy target for the period from 1 January 2015 to 31 December 2017 is set at 700 TWh (lifetime cumulative savings). Since the start of the scheme, 907.4 TWh of energy certificates have been issued, with agriculture accounting for only 5% of the energy certificates issued.

The Agricultural Competitiveness and Adjustment Plan – Reconciliation of economic and ecological performance

The principle of the Competitiveness and Adjustment Plan (PCAE) scheme for agricultural holdings is to provide guidance on to farms on the investments they make (http://agriculture.gouv.fr/sites/minagri/files/plan_pour_la_competitivite_et_ladaptation.pdf). The scheme integrates three programmes: the energy performance plan, the modernisation plan for livestock buildings and the plant plan for the environment. It runs from 2015 to 2020, and covers all agricultural sectors. This scheme is one of the pillars of the Regional Rural Development Plans, which are financed by the central government, the regions and the European Agricultural Fund for Rural Development, with an annual budget of EUR 200 million over the period 2015-20. Funding can also be provided by other sources, such as water agencies, the Agence de l'environnement et de la maîtrise de l'énergie (ADEME) and regional councils.

The focus and objectives of the PCAE have been defined jointly by the central government and the regions and are divided into four axes:

- Livestock, including the poultry and pig sectors: modernisation of livestock buildings, which is the first priority of the plan, in order to reduce energy consumption and develop renewable energy.
- Crops sector: improve the economic and environmental performance of the sector; control of inputs and protection of natural resources (limiting soil erosion, protecting water supplies, encouraging biodiversity); renovation of the orchard sector; and investment in greenhouses, hemp, flax, potato starch and rice.
- Improving the energy performance of all farms: reducing production costs and promoting energy-saving investments and the production of renewable energy on farms, notably through methanisation.
- Cross-cutting priorities: encouraging projects that are part of an agro-ecological approach.

All regions have agreed in their Rural Development Programme to support axis 3 of the PCAE relating to energy performance of farms. Moreover, for 2015 and 2016, the energy-GHG diagnosis remains mandatory to benefit from support on investment related to energy savings, and for the production of renewable energy.

The *Energy Performance Plan* (PPE) for agricultural holdings, launched in 2009, aims to reduce French farms energy use by 30%, through various actions and investments taken after the production of an "energy and GHG emissions audit". The audit's objectives are to: increase awareness of energy consumption on farms by reducing energy consumption, enhancing energy efficiency in agriculture, producing renewable energies, and improving farmers' competitiveness.

The Plan consists of eight pillars:

- Assessment of the energy balance of farms;
- Increased implementation of farm energy and GHG emissions audits (e.g. through the provision of grants);
- Improved tractor-energy efficiency;
- Increased energy efficiency on farms (e.g. through grants offered to farmers who install energy-efficient equipment; encouraging field operations that take into account the reduction of input consumption (such as nitrogen fertilisers), and promoting the use of Energy Performance Certificates);
- Development of renewable energy production, e.g. providing grants to farmers who install renewable energy equipment such as biomass heating systems and solar heating, thermal exchangers and heat pumps, and for methanisation (anaerobic fermentation) units and equipment linked to the production of electricity on an isolated site not connected to the network, such as small wind farms and photovoltaic panels;
- The taking into account of the characteristics of France's overseas territories;
- Promotion of research and innovation; and
- The monitoring and assessment of the PPE.

Ireland: The "Origin Green" commitment to sustainability all along the supply chain

Origin Green is a voluntary sustainability development programme, launched in 2012 by the Irish Food Board. It requires manufacturers to set targets in areas such as emissions, energy, waste, water, biodiversity, and corporate and social responsibility activities. 533 companies have registered with Origin Green, and 224 companies are verified members accounting for 85% of Irish food and drink exports and achieving the target that had been set for December 2016 (www.origingreen.ie/533).

Companies are required to commit to the development and implementation of a sustainability action plan of up to five years' duration, which can be renewed or updated as appropriate at the end of the period. This action plan must clearly set out stretch targets within the key action areas (i.e. sourcing of raw materials, the manufacturing process and social sustainability) identified by the company. For each target area, each company will need to sets out a baseline, decides on short, medium and long term targets and commits to reporting progress on an annual basis. An independent third party evaluates and monitors plans.

The Quality Audit System on farm has been expanded to incorporate sustainability measures to link primary production to food manufacturing. To date sustainability metrics are being measured on 60 000 farms on an 18-month cycle. A third level has been introduced in 2016, with the pilot participation in Origin Green by third party distribution in Ireland as Origin Green moves towards full food chain participation.

New Zealand: Enhancing business growth and competitiveness through energy efficiency improvements

Reducing GHG emissions under its Kyoto Protocol obligation is an additional driver for improving energy efficiency in the food chain. This has led to: setting a target to raise electricity generation from renewables to reach 90% of all electricity generated by 2020 (from 75%); state-owned electricity companies advising farmers on reducing electricity demand; and the launching in 2008 of the NZ Emissions Trading

Scheme (ETS), and has since continued to evolve to cover all sectors of the economy, including agriculture – which currently only has reporting obligations.

In 1992, the New Zealand Energy Efficiency and Conservation Authority (EECA) was established to provide advice and support to businesses, home-owners and farmers on using energy efficiently (www.eeca.govt.nz). By law, there must always be a five-year national Energy Efficiency and Conservation Strategy (NZEECS) to guide EECA's work on energy efficiency, conservation and renewable energy. The plan works in tandem with the New Zealand Energy Strategy, which is overseen by the Ministry of Business, Innovation and Employment. The New Zealand Energy Strategy 2011-21 is the government's ten-year plan for the energy sector and the role energy will play in the country's economy www.eeca.govt.nz/assets/Resources-EECA/nz-energy-strategy-2011.pdf.

The government's overall role in energy efficiency is to provide incentives and information, and to help remove barriers to markets operating effectively. Government support to energy efficiency initiatives for agro-food sector comprise measures such as energy audits, support for energy efficient purchasing, grant and subsidy programmes, and building sector capacity and capability in energy management. The government, in particular, encourages development and use of voluntary industry standards to rate building energy performance.

The means by which the government will work with businesses to achieve these goals by:

- Encouraging businesses to factor in operational costs as well as capital costs when investing in assets, as the longer term energy savings may be worth a higher upfront cost.
- Building management capability, including in small and medium enterprises, to identify and exploit opportunities to ensure energy intensity good practice is reflected in mainstream business planning.
- Encouraging major firms proficient in energy efficiency practices to champion good practice across the wider business community.
- Prioritising energy R&D funding to develop renewable energy and demand side management technologies that improve energy security, and efficient and affordable energy use.

These initiatives have resulted in many successes, such as the major milk processing company Fonterra reducing its energy input per tonne of product by 14% and its on-farm GHG emissions by 9% per litre of milk (Ferrier, 2011).

The Netherlands: Business-driven approach to energy efficiency sets ambitious targets

According to the Environmental Act companies and institutions have to take energy efficiency measures if their initial investment is fully recovered in five years or less. This applies to all sectors with the exception of the glasshouse horticulture sector. The conclusion of a review of the food processing industry is that periodical reviews of energy usage in the production process achieve the desired results.

Examples of success include:

- Primary sector: pre-cooling of milk, insulation of pipes, efficient fans/ventilators, automatic light switches, using body heat of sows for piglets, multilayer production of flower bulbs/mushrooms.
- Food processing industry: energy management systems, reduction in packaging, improved building insulation, dimensioning/seizing the production process, use of up-to-date technologies, and using residual heat, pre-cooling or heating. Each company will need measures specific to its production processes, but improved insulation and energy-efficient lighting are largely applicable across the board.
- The margarine, fats and oil sector: optimising process water, insulation of pipes and appendages.

Long-Term Agreements on energy efficiency

Since the early 1990s, the Government has made voluntary – but not without obligations – long-term agreements on energy efficiency (or covenants) (LTAs) with various industrial and non-industrial sectors an important part of Dutch energy policy. The Long-term Agreements contribute to attaining the 20% CO_2 reduction-target in 2020. At the same time, the aim is to meet energy efficiency targets as set in the Energy Agreement. The time span of the current LTAs varies, but they all end in 2020.

Medium-sized – and sometimes smaller – industrial enterprises take part in LTAs. Larger energy intensive companies participate in the Long-term Agreement on Energy Efficiency for ETS enterprises, that is to say, enterprises that participate in the EU Emissions Trading System. The Agreements are signed by the central government (ministers for Economic Affairs; Infrastructure and the Environment; and Finance; and in the case of LTAs also the minister for the Interior and Kingdom relations), the provincial authorities, the participating companies and relevant trade organisations. Over 40 sectors have signed these agreements, involving over 1 000 companies and representing about 90% of industrial primary energy consumption in the country.

Every four years, participating companies must draft an Energy Efficiency Plan (EEP) mapping out the company's energy efficiency goals, the measures they intend to employ, and a schedule for reaching the goals. An EEP describes measures for improving energy efficiency not only within the company's production process, but it also covers energy management and product and supply chain efficiency. Concerning agriculture, LTAs are in place to improve energy efficiency for the horticulture sector with heated greenhouses.

The total use of energy within the total life cycle of a product, from raw materials up to disposal, is taken into account. Improvements in energy efficiency per sector can result from energy efficiency measures taken by companies to improve the performance of products (process efficiency), and measures taken by companies regarding product and supply chain efficiency. This includes measures such as more efficient transportation, savings in the use phase (e.g. lower energy consumption, lifetime extension), or savings resulting from efficient and effective disposal of products (e.g. re-use, recycling/up-cycling). Companies also report the use of renewable energy.

In terms of monitoring, the LTA programme is implemented by the Netherlands Enterprise Agency, which is part of the Ministry of Economic Affairs and implements government policy for sustainability, innovation, and international business and co-operation. Companies must provide the Netherlands Enterprise Agency with monitoring data, on an annual basis. This information – on the progress they have made with implementing their EEP and the practice of systematic energy management – provides the basis for the sector reports that are discussed each year with the members of the relevant Consultative Group on Energy Conservation of the sector.

Clean and Efficient Covenant for the Agricultural Sectors

The Clean and Efficient Covenant for the Agricultural Sectors 2008-20 (Agro-covenant) is an agreement aimed at achieving the following targets: i) reduction of non-CO_2 GHGs by 4 to 6 mt CO_2-eq by 2020 compared with 1990; ii) an increase of the share of renewable energy to 20% by 2020; and iii) the achievement of an energy efficiency level improvement rate of 2% per year by 2020. The Covenant came into force in June 2008 and will run until December 2020. Considerable attention is given to the efficient use of heat, as well to the production of extra wind and solar power, on land and rooftops of agricultural buildings.

A mix of policy instruments is used to achieve these objectives:

- Funding of research; dissemination of knowledge;
- Communication as to stimulate farmers to incorporate energy efficiency measures in their investments;

- Instruments to stimulate innovation; and
- Normative measures for energy efficiency, CO_2 emissions and sustainability.

The Agro-covenant contains measures for all agricultural sectors, including the food and drink industry. Around 200 small- and medium-sized businesses in the dairy, meat processing, margarine, oils and fats, coffee-roasting, fruit and vegetable processing, cocoa, potato processing and flour milling sectors participate in a separate covenant, "LTAs energy efficiency". Under this agreement, participants endeavour to achieve (on average) for the combined businesses a 30% energy efficiency improvement in the period 2005-20.

The greenhouse horticulture sector used to be part of the Agro-covenant, but agreed upon a new LTA with the government. This energy transition agreement 2014-20 between the greenhouse sector and the Ministry of Economic Affairs is focused on absolute reduction of CO_2-emission by stimulating energy saving and the use of renewable energy. The ambition is a climate neutral greenhouse sector in 2050.

United Kingdom: Climate Change Agreements (CCA) – voluntary industry-led approach to reducing emissions and improving energy efficiency

Climate Change Agreements (CCAs) are voluntary agreements that allow eligible energy-intensive sectors to receive an up to 90% reduction in the Climate Change Levy – a tax on energy consumption – in return for meeting demanding energy efficiency targets agreed with government. CCAs are part of an overall strategy to meet UK Carbon Budgets so their emissions reduction is also important.

The current CCA scheme, which came into effect in 2013, provides an extension to the Climate Change Levy rebate for energy intensive industries until 2023 in return for meeting energy efficiency improvement targets. A total of 53 industrial sectors – including steel, aerospace and farming (i.e. intensive pig and poultry) – have signed up across more than 9 000 locations. Targets apply to participating sectors from 2013 to 2020. According to the Government, the agreement will deliver an 11% energy efficiency improvement across all industry sectors by 2020 against agreed baselines. These savings will be delivered through the implementation of cost-effective measures such as high efficiency motors, variable speed drivers, energy efficient boilers, improved energy management systems and process optimisation. Government is required to review targets set under CCA every 7 years, in order to ensure that targets reflect the full potential for energy efficiency improvements or carbon savings taking into account of any changes in technical or market circumstances.

For the agro-food sectors signed up, the final energy efficiency improvement targets for 2020 (from agreed sector baseline) are: agricultural supply, 7.5%; dairy industry, 13.6%; egg processing, 20%; bakers, 7%; brewers, 13.6%; food and drink, 18%; food and drink – supermarkets, 14%; food storage and distribution federation, 11.7%; horticulture, 14%; meat, 15%; pigs, 22.7%; poultry meat processing, 15%; and poultry meat rearing, 13%. It is estimated that CCAs in pigs, poultry meat, eggs and covered horticulture achieved energy savings of up to 40% have been achieved when compared to their base year energy use in 2000/2001 (DEFRA, 2013a).

The CCA programme has been controversial (Bowen and Rydge, 2011). Some claimed that that the Agreements were effective in focusing managerial attention on energy efficiency, while others have cast serious doubt on their efficacy. In particular, it was argued that they have not been very demanding, and that negotiating and monitoring the Agreements was also a resource-intensive process. In contrast, empirical studies also showed that the full Levy, but not the Agreement, was successful in promoting energy efficiency and innovation, suggesting that there is a case for the abolition of the latter.

United States: A set of voluntary programmes aimed at enhancing energy efficiency

The United States employs a suite of programmes aimed at enhancing energy efficiency. Programmes that encourage energy efficiency and are administered by the USDA include programmes authorised under the Energy Title of the Farm Bill as well as under Conservation and Rural Development Titles, including the

Rural Energy for America Program (REAP), the Multi-Family Housing Energy Initiative, the Rural Utility Service Electric Program, the High Energy Cost Grant program and the Environmental Quality Incentive Program (EQIP) (Farley, 2013; www.ers.usda.gov/agricultural-act-of-2014-highlights-and-implications/).

In general, these programmes support investments in alternative energy technology and the production of renewable biomass for biofuels through education, research and financial assistance programmes; and encourage the manufacture and production of other renewable biochemical and biobased products through federal procurement and financial assistance programmes. Below, three programmes are discussed: The Rural Energy for America Program (REAP), EQIP and the Water and Waste Disposal Program.

The Rural Energy for America Program

The REAP provides financial assistance in the form of loan guarantees and grants to agricultural producers and small businesses to promote energy efficiency and renewable energy development in rural areas. REAP funding is available in several areas: energy audits, energy efficient improvements to farm equipment, and renewable energy projects.

Following amendments in the 2014 Agricultural Act (Farm Bill), the REAP now includes a three-tiered loan and grant application process that sorts proposed projects according to the cost of the proposed activity. In addition, REAP can no longer provide funding for feasibility study grants, nor for blender pumps due to the exclusion of retail energy-delivery mechanisms in the modified definition of "renewable energy system". Councils (i.e. non-profit entities or affiliates) are now eligible to apply for energy audit and renewable energy development assistance grants. REAP expenditure has declined since funding peaked at USD 361 million in fiscal year 2010. Although the programme's funding is reduced to USD 50 million in mandatory funding and USD 20 million in discretionary funding per fiscal year from 2014 through 2018, the programme continues to provide assistance to agricultural producers and small businesses in rural areas for adopting renewable energy and improving energy efficiency.

Environmental Quality Incentive Program

EQIP, which was re-authorised until 2018 in the 2014 Farm Bill, is an important conservation programme, though its importance as a source of energy efficiency funding is sometimes overlooked. EQIP has several initiatives addressing various environmental concerns. It provides financial assistance to producers to install and maintain conservation practices on eligible agricultural and forest land.

Energy conservation and efficiency is only one of the programme's goals. Energy conservation projects occur as part of the On-Farm Energy Initiative. EQIP assists farmers with conducting site-specific energy audits and developing conservation plans, often working through locally-based technical service providers such as EnSave.[13] These energy audits are known as Agricultural Energy Management Plans (AgEMPs).

There are two types of AgEMPs: Headquarters AgEMPs address some of the most traditional farm energy concerns, such as lighting efficiency and fuel use. Landscape AgEMPs have been available since 2009, and take a more holistic view of on-farm energy use, addressing issues, such as water use and erosion. Once the initial energy audit is complete, EQIP helps farmers develop a plan for implementing conservation practices. Funding is available to assist farmers with implementation. EQIP funding is provided on a first-come first-served basis rather than based on a competitive application process.

Water and Waste Disposal Program

The Water and Waste Disposal Program provides financing for rural communities to establish, expand or modernise water treatment and waste disposal facilities. Projects are designed to improve the energy efficiency of the water and waste facilities and to improve water conservation efforts. Eligibility is limited to communities of 10 000 or less in populations that are unable to obtain credit elsewhere. In addition, financing is available only to those communities with low median household income levels. Priority is given to public entities serving areas with populations less than 5 500 and applying for loans to restore a deteriorating water system or to improve, enlarge or modify an inadequate waste facility. Grants are limited to a maximum of 75% of project costs. Programme regulations stipulate that the grant amount should only be as much as necessary to bring the user rates down to a reasonable level for the area. Water and Waste Disposal grant and loan funds are usually combined based on the income levels and user costs. However, separate stand-alone grants are provided for solid waste disposal and technical assistance and training. The 2016 Budget provides a total water and waste disposal programme level of nearly USD 1.7 billion.

Notes

1. Gillingham and Palmer (2013) find that engineering studies may over-estimate the size of the energy efficiency gap by failing to account for all costs and neglecting particular types of economic behaviour. Nonetheless, empirical evidence suggests that market failures, such as asymmetric information and agency problems affect efficiency decisions and contribute to the gap.

2. For example, Galitsky et al. (2003) have identified over 100 technologies and measures for improving energy efficiency in the food processing and manufacturing sector.

3. In a Nestlé factory, it was found that energy savings of up to 30% can be achieved by using procedural and behavioural changes, and without capital investment.

4. Allcot and Greenstone (2012), in their review of the empirical studies on the magnitude of the Energy Efficiency Gap, argue that this literature frequently fails to meet modern standards for providing credible estimates of the net present value of energy-cost savings, and often leaves other benefits and costs un-measured.

5. Recently, some economists have proposed that systematic behavioural biases in consumer decision-making may explain the apparent efficiency gap. See Gillingham and Palmer (2013) for more discussion.

6. This support is hampering global efforts to curb emissions and combat climate change. Governments are spending almost twice as much money on support for fossil fuels as is needed to meet the climate-finance objectives set by the international community, which call for the mobilisation of USD 100 billion a year by 2020 (OECD, 2015a).

7. It should be noted that some studies find no direct causal link between the prices of energy and non-energy commodities (Baffes and Haniotis, 2016). The divergent findings may reflect either the rising importance of biofuels (that may have weakened the link between oil and food commodity prices), or methodological differences.

8. Fertiliser prices are down 45% since 2011 and more than 50% lower since their all-time high in 2008.

9. Changes in production costs are small for two reasons: i) prices for energy-related inputs decline by less than change in prices; and ii) energy-related cost represent only a portion of total operating expenses.

10. Recent policy announcements, including the EU Energy Efficiency Directive and the US Clean Power Plan, will support greater levels of investment, as will the Intended Nationally Determined Contributions (INDCs), submitted to the UN Framework Convention on Climate Change (UNFCCC) over 2014, which entail expanded energy efficiency action.

11. However, in many countries similar campaigns have been successful at addressing non-food-related health issues, such as nicotine addiction and drunk driving. Successes have been attributed to long-term

campaigns that continually reinforce the message of what constitutes good behaviour. In most cases, information has been accompanied by measures to restrict individual freedom – for example, making drink-driving and not wearing seat belts illegal.

12. In Finland, 88% of the energy consumption in agriculture could be produced through biogas, although biogas production is still modest (Huttunen and Kuittinen, 2015).

13. EnSave is a Vermont-based company that assists farmers with energy-efficiency upgrades through energy audits and consultation services. For more information, see www.ensave.com.

Bibliography

Allcott, H. and M. Greenstone (2012), "Is there an energy efficiency gap?", *National Bureau of Economic Research*, Working Paper 17766.

Ambec, S., M. Cohen, S. Elgie and P. Lanoie (2011), "The Porter Hypothesis at 20: Can Environmental Regulation Enhance Innovation and Competitiveness?", *Resources for the Future*, Discussion Paper 11-01, Washington, D.C.

Baffes, J. and T. Haniotis (2016), *What Explains Agricultural Price Movements?*, Policy Research Working Paper 7589, World Bank Group, Washington, D.C.

Baffes, J. M. Ayhan Kose, F. Ohnsorge and M. Stocker (2015), *The Great Plunge in Oil Prices: Causes, Consequences, and Policy Responses*, Policy Research Notes 15/1, World Bank Group, http://pubdocs.worldbank.org/en/339801451407117632/PRN01Mar2015OilPrices.pdf

BASE (2012), *Energy Savings for a Bread Plant*, Base Company, http://baseco.com/casestudies/Bread%20Plant.pdf.

Bio-Based Industries (BBI) (2015), Public-Private partnership website, www.bbi-europe.eu/about/about-bbi.

Business and Industry Advisory Committee to the OECD (BIAC) (2013a), *Green Growth in the Agro-food Chain: What Role for the Private Sector?,* BIAC Discussion Paper, https://www.oecd.org/tad/sustainable-agriculture/Session%201%20KIRCHBERG_Paper%2017%20april%20BIAC%20paper%20formatted.pdf.

BIAC (2013b), "BIAC Perspectives on Private Sector Solutions to Food Waste and Loss", paper prepared for the 4th OECD Food-chain Network Analysis Meeting on Food Waste along the Supply Chain, 20-21 June 2013, www.oecd.org/site/agrfcn/BIAC%20Perspectives%20on%20Private%20Sector%20Solutions%20to%20Food%20Waste%20and%20Loss.pdf

Bowen, A. and J. Rydge (2011), "Climate-Change Policy in the United Kingdom", OECD Economics Department Working Papers, No. 886, OECD Publishing. http://dx.doi.org/10.1787/5kg6qdx6b5q6-en.

British Retail Consortium (BRC) (2014), *A Better Retailing Climate: Driving Resource Efficiency*, www.brc.org.uk/retailingclimate.

British Sugar (2010), *2009-10 Corporate Sustainability Report*, Peterborough, British Sugar, http://britishsugar2.tinfishcreative.co.uk/Files/British-Sugar-Sustainability-Report-NAVIGABLE_2(2).aspx.

Brown, E., R. Elliott and S. Nadel (2005), "Energy efficiency programs in agriculture: design, success, and lessons learned", *American Council for an Energy-Efficient Economy (ACEEE)*, Report No. IE051.

Capacci S., M. Mazzocchi, B. Shankar, J. Brambila Macias, W. Verbeke, F.J.A. Pérez-Cueto, A. Kozakowska, B. Piórecka, B. Niedzwiedzka, D. d'Addesa, A. Saba, A. Turrini, J. Aschemann-Witzel, T. Bech-Larsen, M. Strand, J. Wills and B. Traill(2012), "Policies to promote healthy eating in Europe: A structured review of instruments and their effectiveness", *Nutrition Reviews* , Vol. 70, No. 3.

Department for Environment, Food and Rural Affairs (DEFRA) (2013), *Energy Dependency and Food Chain Security*, Report FO0415, DEFRA, London.

DEFRA (2010), *Food 2030 Strategy*, DEFRA, London.

EATWELL (2012), "Effectiveness of Policy Interventions to Promote Healthy Eating and Recommendations for Future Action: Evidence from the EATWELL Project", Deliverable 5.1, http://eatwellproject.eu/en/upload/Reports/Deliverable%205_1.pdf.

Farley, K. (2013), "Energy efficiency opportunities at USDA: An ACEEE White Paper", American Council for an Energy-Efficient Economy (ACEEE), Washington, D.C.

Energy Efficiency and Conservation Authority (EECA) (2014), "Irrigation energy efficiency evaluation pilot – summary report. New Zealand", www.eeca.govt.nz/resource/irrigation-energyefficiency-evaluation-pilot-summary-report.

EECA (2009), "Dairy farmers milk free energy", www.solarthermalworld.org/sites/gstec/files/New%20Zealand%20Dairy%20Farms.pdf.

FoodDrinkEurope (2012), *Environmental Sustainability Vision Towards 2030*, http://sustainability.fooddrinkeurope.eu/uploads/section-images/USE_SustainabilityReport_LDFINAL_11.6.2012.pdf, accessed September 2015.

Galitsky, C., E. Worrell, A. Radspieler, P. Healy and S. Zechiel (2005), *BEST Winery Guidebook: Benchmarking and Energy and Water Savings Tool for the Wine Industry*, Berkeley, CA: Lawrence Berkeley National Laboratory, LBNL-3184.

Gillingham, K. and K. Palmer (2013), "Bridging the Energy Efficiency Gap Insights for Policy from Economic Theory and Empirical Analysis", *Resources for the Future*, DP 13-02.

Gillingham, K., R. Newell and K. Palmer (2009), "Energy Efficiency Economics and Policy", *National Bureau of Economic Research*, Working Paper 15031.

Gołaszewski, J., C. de Visser, Z. Brodzinski, R. Myhan, E. Olba-Ziety, M. Stolarski, G. Papadakis (2012), "State of the art on energy efficiency in agriculture - Country data on energy consumption in different agro-production sectors in the European countries", Project founded by the FP7 Programme of the EU with the Grant Agreement Number 289139, AGREE Project Deliverable 2.1.), www.agree.aua.gr/.partnersarea/index-3.html.

Hasler, M. (2008), "Health claims in the United States: An aid to the public or a source of confusion?", *The Journal of Nutrition*, Vol. 138, No. 6.

Hausman, C. and R. Kellogg (2015), "Welfare and Distributional Implications of Shale Gas", *Brookings Papers on Economic Activity*, Brookings Institution, Washington, D.C.

Hitaji, C. and S. Suttles (2016*), Trends in U.S. Agriculture's Consumption and Production of Energy: Renewable Power, Shale Energy and Cellulosic Biomass*, EIB-159, Economic Research Service, United States Department of Agriculture, Washington D.C.

Huttunen, M.J. and V. Kuittinen (2015), *Suomen biokaasulaitosrekisteri*, No. 18, Reports and studies in Forestry and Natural Sciences No 21. University of Eastern Finland.

International Energy Agency (IEA) (2015), *World Energy Outlook Special Report on Energy and Climate Change*, IEA/OECD, Paris.

IEA (2014), Capturing the Multiple Benefits of Energy Efficiency, IEA, Paris.

IEA (2008), *Energy Efficiency Policy Recommendations Prepared by the IEA for the G8 Under the Gleneagles Plan of Action*, IEA/OECD, Paris.

Lang, T. (2013). "Lessons from Across the Atlantic: Policy Faultiness and Policy Possibilities on Sustainable Diets", available at http://iom.edu/~/media/Files/Activity%20Files/Nutrition/FoodForum/May_07_2013/Tim%20Lang.pdf.

Marshall, K., S. Riche, R. Seeley and P. Westcott (2015), *Effects of Recent Energy Price Reductions on U.S. Agriculture*, Economic Research Service, United States Department of Agriculture, Bio-04, Washington, D.C.

McKinsey & Company (2010), *Energy efficiency: A compelling global resource.*

Mehlhart, G., I. Bakas, M. Herczeg. P. Strosser, C. Rynikiewicz, A. Agenais, T. Bergmann, M. Mottschall, A. Köhler, F. Antony, V. Bilsen, S. Greeven, P. Debergh and D. Hay (2016), *Study on the Energy Saving Potential of Increasing Resource Efficiency – Final Report*, Study prepared for the European Commission, Directorate General Environment, Brussels, http://ec.europa.eu/environment/enveco/resource_efficiency/pdf/final_report.pdf

Mhurchu, C. and D. Gordon (2007), "Nutrition labels and claims in New Zealand and Australia: a review of use and understanding", *Australian and New Zealand Journal of Public Health*, Vol. 31, No. 2.

Morvaj, Z. and V. Bukarica (2010), "Energy efficiency policy", in Palm, J. (ed.) (2010), *Energy Efficiency*, https://cdn.intechopen.com/pdfs-wm/11461.pdf.

Nadel, S. and H. Geller (1996), "Utility DSM: What Have we Learned? Where are we Going?", *Energy Policy*, Vol. 24, No. 4.

OECD (2016), "Synergies and Trade-offs between Agricultural Productivity and Climate Change Mitigation and Adaptation: Dutch Case Study", COM/TAD/CA/ENV/EPOC(2016)7/FINAL, OECD Publishing, Paris.

OECD (2015), *Taxing Energy Use 2015: OECD and Selected Partners Economies*, OECD Publishing, Paris, doi:10.1787/9789264232334-en

OECD (2013), *Effective Carbon Prices*, OECD Publishing, Paris, http://dx.doi.org/10.1787/9789264196964-en.

OECD (2014), *Green Growth Indicators for Agriculture – A Preliminary Assessment*, OECD Green Growth Studies, OECD Publishing, Paris, doi:10.1787/9789264223202-en.

OECD (2010), *Taxation, Innovation and the Environment*, OECD Publishing, Paris, doi:10.1787/9789264087637-en.

OECD (2006), *The Political Economy of Environmentally Related Taxes*, OECD Publishing, Paris, doi:10.1787/9789264025530-en.

OECD (2001), *Environmentally Related Taxes in OECD Countries: Issues and Strategies*, OECD Publishing, Paris.

Passey, R. and R. MacGill (2009), "Energy Sales Targets: An alternative to White Certificate Schemes", *Energy Policy*, Vol. 37, No. 6.

PEF World Forum Newsletter (2014), www.pef-world-forum.org/eu-environmental-footprinting/position-papers/ accessed 5 January 2015.

Pollard, M., R. Miller, M. Daly, E. Crouchley, J. O'Donoghue, J. Lang and W. Binns (2008), "Increasing fruit and vegetable consumption: success of the Western Australian Go for 2&5 campaign", *Public Health Nutrition*, Vol. 11, No. 3.

Porter, M. and C. van der Linde (1995), "Toward a New Conception of the Environment-Competitiveness Relationship", *The Journal of Economic Perspectives*, Vol. 9, No. 4.

Sands, R. and P. Westcott (eds.) (2011), *Impacts of Higher Energy Prices on Agriculture and Rural Economies*, Economic Research Service, United States Department of Agriculture, ERS Report No. 123.

Segerson, K. and T. Miceli (1998), "Voluntary Environmental Agreements: Good or Bad News for Environmental Protection?", *Journal of Environmental Economics and Management*, Vol. 36, No. 2.

Sorrell, S., A. Mallett and S. Nye (2011) "Barriers to industrial energy efficiency: A literature review"; UNIDO, 2011.

Tassou, S., M. Kolokotroni, B. Gowreesunker, V. Stojkeska, A. Azapagic, P. Fryer and S. Bakalis (2014), "Energy Demand and Reduction Opportunities in the UK Food Chain", *Energy*, Vol. 167, No. 3.

Vyth, E., I. Steenhuis, A. Roodenburg, J. Brug and J. Seidell1 (2010), "Front-of-pack nutrition label stimulates healthier product development: a quantitative analysis", *The International Journal of Behavioral Nutrition and Physical Activity*, Vol. 7, No. 65.

Ziesemer, J. (2007), *Energy Use in Organic Food Systems*, FAO, Rome.

Annex 5A.1

Selected energy efficiency options of the food chain

Table A5.1. Energy efficiency options of areas of the food chain that are particularly energy-intensive

Production/consumption	Energy demand	Energy efficiency option
Inputs		
Nutrients	Fertiliser use	Optimising the use of nutrients by precision dosing and timing of fertilisers applications
Machinery	Tractor and machinery performance	Using fuel efficient tractors; training machinery operators
On-farm animal feed production (from grazing and crops)	Fertiliser use	Precision application; organic fertilisers
Irrigation	Electricity	Precision irrigation; proper pump/motor sizing according to water demands; GPS sprinkler controls
Primary production		
Arable crops	High fuel (diesel) demand for machinery; high energy demand if irrigated; electricity for irrigation, storage facilities; heat for drying (LPG, gas).	Targeted use of agro-chemicals and manure by farmers; reduce soil tillage practices; use precision farming techniques; organic farming
Vegetables - large scale for processing	High fuel (diesel) demand for machinery; high power demand if irrigated and for post-harvest chillers	Advanced air circulation fan designs; combined heat and power (CHP); heat recovery; heat pumps
Dairy (large scale > 50 cows)	High demand for electricity for milking, pumping, cooling, irrigation, lighting, pasteurising	Optimising pre-cooling; biomass boilers; high efficiency lighting
Intensive livestock	High energy use if mainly housed indoors and feed grown; medium to low if partly outdoors and if feed bought	Manure for biogas; installing biomass boilers; using high efficiency lighting; improving insulation and ventilation systems for livestock buildings
Processing	Heat and cooling	Re-circulation of air in dryer (vegetables); recover heat for pre-heating; pre-cooling methods
Retailing	Storage; ventilation and air conditioning; lighting	Increase efficiency of the refrigeration system; combined heat and power (CHP); heat recovery, heat pumps
Packaging	Packaging materials	Innovation in environmentally-friendly packaging materials and in design; recycling of packaging materials; use bio-based resources (eco-design)
Transport	Diesel fuel use	Adopting an integrated approach that balances transport mode and distance; driver training
Households	Increasing consumer demand for more processed, convenience and take-away foods that require more energy	Minimise food waste; optimal use of domestic appliances (e.g. electricity and other fuels used for cooking, cleaning and food storage)

ORGANISATION FOR ECONOMIC CO-OPERATION AND DEVELOPMENT

The OECD is a unique forum where governments work together to address the economic, social and environmental challenges of globalisation. The OECD is also at the forefront of efforts to understand and to help governments respond to new developments and concerns, such as corporate governance, the information economy and the challenges of an ageing population. The Organisation provides a setting where governments can compare policy experiences, seek answers to common problems, identify good practice and work to co-ordinate domestic and international policies.

The OECD member countries are: Australia, Austria, Belgium, Canada, Chile, the Czech Republic, Denmark, Estonia, Finland, France, Germany, Greece, Hungary, Iceland, Ireland, Israel, Italy, Japan, Korea, Latvia, Luxembourg, Mexico, the Netherlands, New Zealand, Norway, Poland, Portugal, the Slovak Republic, Slovenia, Spain, Sweden, Switzerland, Turkey, the United Kingdom and the United States. The European Union takes part in the work of the OECD.

OECD Publishing disseminates widely the results of the Organisation's statistics gathering and research on economic, social and environmental issues, as well as the conventions, guidelines and standards agreed by its members.

OECD PUBLISHING, 2, rue André-Pascal, 75775 PARIS CEDEX 16
(51 2017 08 1 P) ISBN 978-92-64-27852-3 – 2017

www.ingramcontent.com/pod-product-compliance
Lightning Source LLC
LaVergne TN
LVHW081414110826
845149LV00010B/1740

* 9 7 8 9 2 6 4 2 7 8 5 2 3 *